AF621542

01-WSM-2a

SENSOR TECHNOLOGY FOR WATER QUALITY MONITORING: BIOLUMINESCENT MICROORGANISMS

by:

Paul Frymier, Ph.D.
Shijin Ren, Ph.D.
University of Tennessee and the
Center for Environmental Biotechnology

Curtis Lajoie, Ph.D.
Christine Kelly, Ph.D.
Syracuse University

2004

WERF

The Water Environment Research Foundation, a not-for-profit organization, funds and manages water quality research for its subscribers through a diverse public-private partnership between municipal utilities, corporations, academia, industry, and the federal government. WERF subscribers include municipal and regional water and wastewater utilities, industrial corporations, environmental engineering firms, and others that share a commitment to cost-effective water quality solutions. WERF is dedicated to advancing science and technology addressing water quality issues as they impact water resources, the atmosphere, the land, and quality of life.

For more information, contact:
Water Environment Research Foundation
635 Slaters Lane, Suite 300
Alexandria, VA 22314-1177
Tel: (703) 684-2470
Fax: (703) 299-0742
www.werf.org
werf@werf.org

This report was co-published by the following organizations. For nonsubscriber sales information, contact:

IWA Publishing
Alliance House, 12 Caxton Street
London SW1H 0QS, United Kingdom
Tel: +44 (0) 20 7654 5500
Fax: +44 (0) 20 7654 5555
www.iwapublishing.com
publications@iwap.co.uk

Library of Congress Catalog Card Number: 2003108205

IWAP ISBN: 1-84339-714-5

This report was prepared by the organization(s) named below as an account of work sponsored by the Water Environment Research Foundation (WERF). Neither WERF, members of WERF, the organization(s) named below, nor any person acting on their behalf: (a) makes any warranty, express or implied, with respect to the use of any information, apparatus, method, or process disclosed in this report or that such use may not infringe on privately owned rights; or (b) assumes any liabilities with respect to the use of, or for damages resulting from the use of, any information, apparatus, method, or process disclosed in this report.

University of Tennessee and the Center for Environmental Biotechnology, and Syracuse University
This document was reviewed by a panel of independent experts selected by WERF. Mention of trade names or commercial products does not constitute WERF nor EPA endorsement or recommendations for use. Similarly, omission of products or trade names indicates nothing concerning WERF's or EPA's positions regarding product effectiveness or applicability.

The research on which this report is based was funded, in part, by the United States Environmental Protection Agency through Cooperative Agreement No. CR-827345-01 with the Water Environment Research Foundation (WERF). Unless an EPA logo appears on the cover, this report is a publication of WERF, not EPA. Funds awarded under the Cooperative Agreement cited above were not used for editorial services, reproduction, printing, or distribution.

ACKNOWLEDGMENTS

In addition to the project subcommittee members listed below, the project team members would like to thank Debbie Ailey, David Obenschain, and Monica Burger of the Knoxville Utilities Board for their advice and for their willingness to share data and samples for this project. The project team would also like to thank the Center for Environmental Biotechnology at the University of Tennessee for its financial assistance with cost-sharing on this project.

Report Preparation

Principal Investigator:
Paul Frymier, Ph.D.
University of Tennessee and the Center for Environmental Biotechnology

Project Team:
Shijin Ren, Ph.D.
University of Tennessee and the Center for Environmental Biotechnology

Curtis Lajoie, Ph.D.
Christine Kelly, Ph.D.
Syracuse University

Project Subcommittee

Robin Autenrieth, Ph.D.
Texas A&M University

Donald Brown
U.S. Environmental Protection Agency

William J. Davies
U.S. Geological Survey

James Dartez
Royce Technologies

Bennett K. Horenstein
East Bay Municipal Utility District

Darrell Jan, Ph.D.
Jet Propulsion Laboratory

Jim Lazorchak
U.S. Environmental Protection Agency

Water Environment Research Foundation Staff

Director of Research:	Daniel M. Woltering, Ph.D.
Project Manager:	Dean E. Carpenter, D. Eng.
	Eugenia Bruno Hart (from August 2003)

ABSTRACT AND BENEFITS

Abstract:

Biological wastewater treatment plants can be adversely affected by influent toxicity. The effects can range from poor clarifier biomass settling and elevated effluent BOD (biological oxygen demand) and ammonia levels to total plant kills. These problems could be minimized or eliminated if an effective method existed for continuously monitoring biological wastewater treatment plant influent for toxicity to the treatment plant microorganisms.

Current influent screening methods have not been proven to be adequate for adaptation to continuous screening in the field. The primary reasons include the batch-wise nature of the assays and an inadequate correlation between the assays and plant performance. The goal of the research team was to create new bioluminescent biosensors from different types of bacteria found in biological wastewater treatment plants to develop a multi-channel continuous monitoring system. A system built from multiple biosensors would make it possible to differentiate between potential influent toxicity effects on different classes of bacteria (such as nitrifying and heterotrophic bacteria).

The research team found it unexpectedly challenging to apply common microbiological transformation methods for laboratory strains to the wastewater treatment plant strains. The team generated six new bioluminescent bioreporters from bacteria that are typical constituents of activated sludge. Of particular significance is a bioreporter developed using a *Hyphomicrobium sp.*, which is a slow-growing bacterium known to be present in significant numbers in some activated sludge plants. Of the six strains generated, initial bioluminescence and toxicity screening indicated that one (a *Pseudomonad*) was a particularly promising candidate due to its ease of cultivation and high light production. Further toxicity testing, however, determined that the response of the strain to 48 organic compounds and eight metals commonly found in wastewater was similar to that of a previously created strain, Shk1 (also a *Pseudomonad*). Further work is therefore needed in the generation of appropriate biosensors and test conditions for populations not represented by the new heterotrophic biosensor.

Benefits:

- The report details methodologies appropriate for the creation of new bioluminescent biosensors from bacteria that are common constituents of activated sludge and discusses potential pitfalls for future investigators.
- Six new biosensors were created from common wastewater treatment plant bacteria.
- The methodology details the importance of preliminary toxicity screening in the selection of the components of a test battery for influent toxicity screening assays.

Keywords: Wastewater treatment, biosensors, bacteria, toxicity monitoring, influent screening

TABLE OF CONTENTS

LIST OF TABLES

LIST OF FIGURES

EXECUTIVE SUMMARY

This document describes efforts to generate new biosensor microorganisms for screening influent wastewater for toxicity to biological wastewater treatment systems. From the results of previous WERF-sponsored research (98-CTS-6RD), the project team concluded that a potentially valuable research direction for developing an effective influent toxicity monitoring system would be to construct an array of biosensor organisms including representative species from the different important groups of microorganisms in a biological wastewater treatment plant. The earlier work indicated that, although well correlated with respirometry data, the response of a single biosensor did not accurately predict several key measures of plant performance (like ammonia and COD removal) when used to sample the influent of a bench-scale wastewater treatment plant periodically spiked with high levels of heavy metals. Therefore, the project team proposed generating additional biosensor organisms that would be included in a multi-channel system whose composite signal would be used to attempt to better predict the effect of toxicity on treatment plant performance.

ES.1 Characteristics of Effective Influent Toxicity Biosensors

In order for a monitoring system to effectively screen wastewater treatment plant influent for toxicity, it should possess certain characteristics:

- The system should generate a warning signal at toxicity levels appropriate for the system it is designed to protect. (It should be neither too sensitive nor too insensitive.)
- The system should alert plant personnel in a timely fashion so that protective action can be taken.
- The signal generated by the system should give some indication of the appropriate type of protective action.
- The system should operate without constant operator attention and under less-than-ideal conditions.

Several methods are currently used for toxicity screening of influent wastewater, and most operators of biological treatment plants have an established toxicity-screening program. The primary drawbacks of these programs are that the sampling frequency is insufficient and samples are not tested quickly enough for the program to serve a protective function. The function of most programs is therefore merely diagnostic; a problem is seen in the wastewater treatment plant, and the data from influent toxicity screening efforts are analyzed in retrospect to determine if influent toxicity is a likely cause of the problem.

There are many reasons that toxicity screening is applied in this fashion. One reason is that the assays performed are not of the appropriate sensitivity for determining the potential toxicity of influent to wastewater treatment microorganisms. Most commercially available bioassays were designed to detect toxicity in effluent streams where a highly sensitive assay for toxicity is desired, and a false negative is much more problematic than a false positive. Another reason is that most easily applied bioassays are batch in nature and require the manual collecting and processing of samples. While the methodology of commercially available assays has been designed to make the testing of samples as simple as possible, the batch-wise nature of the assays and the dependence on specialized equipment frequently relegate sample testing to laboratory personnel removed from the collection site. The result is frequently that samples are collected on one day and tested on another. For the purpose of taking protective action, this is highly undesirable. Lastly, most toxicity screens suffer from the inability to determine what part of the treatment process in a multi-stage process is likely to be affected. Not only are some popular commercially available systems excessively sensitive, but plant personnel often resort to a perfunctory testing of the influent with no idea of how to react appropriately in the event of a positive signal.

ES.2 Research Approach

Having considered the abovementioned inadequacies of current systems as well as previous research on the development of toxicity monitoring systems, the project team felt that the most effective system should have the following design characteristics:

- The system should be based on bioluminescent biosensors. These biosensors have already gained acceptance in the water quality monitoring community. Light is a signal that is easily measured, unlike respiration rate, which has to be inferred from the derivative of the oxygen concentration in the aqueous phase.
- The biosensors should be based on native wastewater treatment plant microorganisms. Native microorganisms are more likely to generate responses proportional to the degree to which toxicity will affect treatment plant performance.
- The genes for light production in the biosensors should be under the control of constitutive promoters. Biosensors have been developed with the genes for light production under the control of so-called "shock promoters". These biosensors are effective for signaling specific conditions, but toxicity is a general quality and elicits a composite response from microorganisms. In addition, different mechanisms of action exist for different classes of toxicants so a bioreporter for one specific response would be a poor reporter for another.
- Multiple biosensors generating multiple channels of data are more likely to produce useful data than would a single biosensor. For example, a biosensor from a heterotrophic population may not accurately predict damage to nitrifying populations, and vice-versa.
- Each of the biosensors to be used in the multi-channel system should provide accurate and complementary or unique information. The biosensors should be carefully screened to eliminate data redundancy.
- The system should allow for easy adaptation to a continuous monitoring system. Commercially available batch-wise systems have been developed over the years

to require no significant advance preparation in order to perform an assay. However, they suffer from the constraint that they can, at best, be implemented in a semi-batch fashion for continuous monitoring.

In order to achieve their goal of a multi-channel bioreporter-based continuous monitoring system, the research team's initial goal was to create new biosensors (in addition to one previously created) by starting either with bacteria isolated from activated sludge or with bacteria purchased from a culture collection from species known to be present in biological wastewater treatment systems. To that end, the team added the genes necessary to give the bacteria the ability to produce light to each of the original wastewater treatment strains collected by the appropriate transformation vector. Then the newly created bioluminescent bacteria were carefully screened to determine if they produced adequate light to generate a useful signal and which of the biosensors would be useful to include in a multi-channel system without generating redundant information.

ES.3 Creation of Biosensors and Toxicity Testing

Creating the new biosensors was unexpectedly challenging. Transformation techniques known to be effective with typical laboratory bacteria (like *Escherichia coli*) were much less effective with many typical wastewater treatment strains. Techniques used included matings of pure cultures with transformed *E. coli*, electroporation of a self-replicating plasmid, electroporation of a linearized transposon vector for chromosomal insertion, among others. Establishing appropriate counter-selection methods was also challenging. Although biosensors made from wastewater treatment bacteria (particularly nitrifiers) were known to have been created in two other laboratories, further communication with the laboratories indicated that the strains had either not been maintained (at one laboratory) or were held by a company laboratory, which complicated their transfer. Six new biosensor organisms—three *Hyphomicrobium* and three *Pseudomonas* species—were successfully created.

The newly created bioluminescent strains were then subjected to preliminary tests to see if they produced adequate light for further toxicity testing. First, team members at Syracuse University employed a preliminary screening assay that used synthetic wastewater or actual wastewater (from a wastewater treatment plant in Syracuse, New York) as the dilution medium into which the toxicants and bioreporters were added. These early tests indicated that the six strains produced sufficient light to warrant further screening. The strains were then transferred to the project team members at the University of Tennessee (UT). The research team conducted additional preliminary studies designed to screen for the most promising strains and determined that five of the six new strains produced levels of light that were too low to get accurate toxicity data. The team members at UT employed a batch-wise toxicity testing protocol that used distilled water as the dilution liquid for the toxicants and bioreporters. The remaining strain (a *Pseudomonad* designated PM6) grew quickly and produced ample light; this strain was tested with 48 organic compounds and eight metals that are frequent constituents of industrial and domestic wastewater. The toxicity data were compared to other bioluminescence-based toxicity assays and to activated sludge inhibition data in the literature.

ES.4 Conclusions

The research team successfully generated six new bioluminescent biosensors. Of the six, toxicity testing showed that one (the *Pseudomonad* designated PM6) was a particularly good candidate for more extensive testing due to its high light level and growth rate. In addition, preliminary data gathered at Syracuse University for the *Hyphomicrobium* spp. biosensors indicated that they did produce measureable light and had the advantage of representing a different genus of bacterium than the *Pseudomonads*.

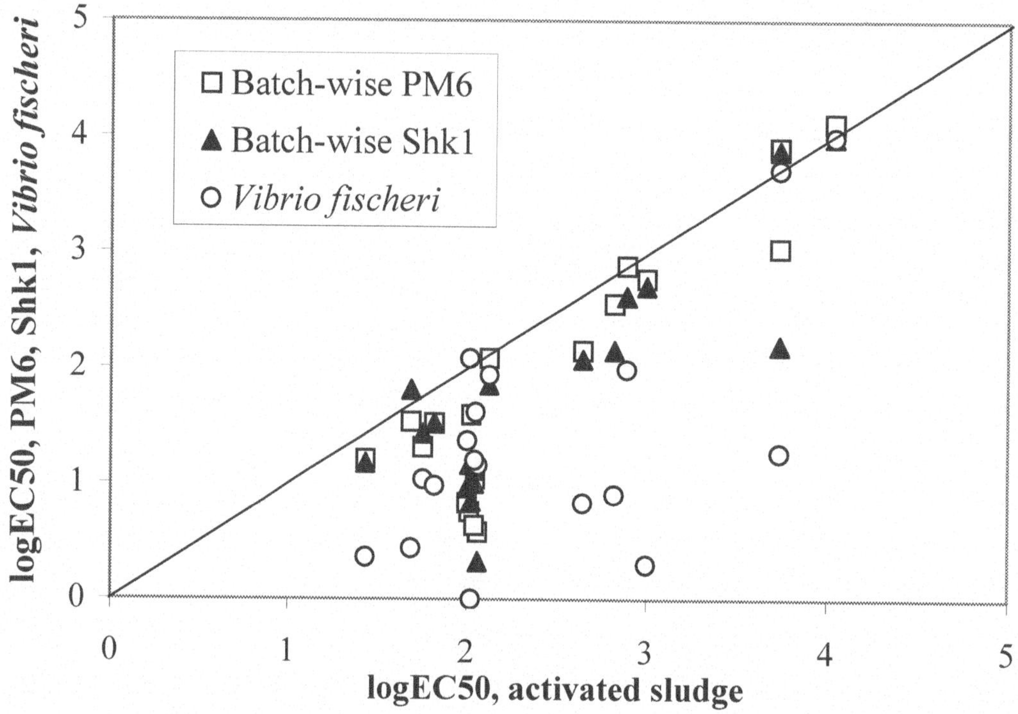

Figure ES-1. Comparison of EC$_{50}$ Values for Three Bioluminescent Toxicity Assays to EC50 Values for Activated Sludge Respirometry Inhibition Responding to Metals and Organic Compounds. The Data for *Vibrio fischeri* are taken from the Literature.

Their low growth rate (batch culture times on the order of a week) and lower light level led the researchers to eliminate the *Hyphomicrobium* biosensors from further screening at this time; the project team concluded that a biosensor with this low growth rate would be very difficult to maintain in continuous culture for a continuous monitoring system. The team continued detailed testing with the *Pseudomonad* PM6. This more detailed toxicity testing revealed that the new biosensor's response to the test compounds was highly correlated to that of a previously created biosensor, Shk1 (also a

Pseudomonad). On a positive note, these two biosensors (PM6 and Shk1) exhibit a stronger correlation with activated sludge respirometry toxicity data than a commercial assay based on the marine organism *Vibrio fischeri*. (See Figure ES-1; when the values of each assay are correlated with the activated sludge data, the resulting R^2 values for the correlation of each assay are $R^2_{PM6}=0.82$, $R^2_{Shk1}=0.76$, $R^2_{V. fischeri}=0.43$) However, the goal of the research team was to create new biosensors that would generate unique information. The team concluded that the strong correlation between Shk1 and PM6 indicated that, while either sensor would be a useful surrogate for direct measurements of respiration inhibition in activated sludge, their signals are not unique, so if both were used in a test battery, one of the assays would be redundant.

ES.5 Recommendations for Future Research

Based on the results of this project, the research team identified several potential areas for future research in the short and long term and some suggestions to future researchers in this area.

Short-term research should focus on two tasks: 1) obtaining a nitrifying biosensor either by more focused effort to insert the available plasmid (Iizumi, T., M. Mizumoto, and K. Nakamura. Appl. Environ. Microbiol. 64(10): 3656-3662, 1998) in a *Nitrosomonas* bacterium or by obtaining the biosensor from the only group known to have maintained a bioluminescent nitrifying bacterium (Ludwig C, S. Ecker, K. Schwindel, H.G. Rast, K.O. Stetter, and G. Eberz, Archives of Microbiology, 172 (1): 45-50, 1999), and 2) developing an assay that has the sensitivity and ability to accommodate the low growth rate required to utilize the *Hyphomicrobium* biosensor developed as part of this project. With these two organisms, in addition to PM6 or Shk1, it would be possible to test the central hypothesis of the original project; that is, additional biosensors can be created that give unique information about the toxicity of a sample.

Long-term research should focus on developing both more effective methods of culturing populations of slow-growing wastewater treatment plant bacteria and more effective methods of getting the *lux* genes into these bacteria. Many wastewater treatment plant bacteria are slow-growing microorganisms that do not grow well or cannot be cultured using current standard laboratory techniques. Simply using spread plate and conjugation methods with a consortium of wastewater treatment plant bacteria and *E. coli* carrying the *lux* genes on a self-replicating plasmid will therefore result most frequently in bioreporters that are fast-growing microorganisms present in high numbers, like *Pseudomonads*. This is why pure cultures were used in the transformation efforts documented in this report. The pure culture transformation efforts of the research team further indicate that, unlike *Pseudomonads*, some wastewater treatment plant microorganisms cannot maintain the self-replicating plasmid pUTK2 and must be transformed with transposon vectors that allow the genes to be inserted in the chromosomal DNA. Many wastewater treatment plant bacteria have protective sheaths on their filaments or exopolysaccharide layers on their exteriors that the research team has observed makes it harder to get genes across the membrane of the target cells. In earlier work by another researcher (data not shown), chemical mutagenesis was necessary to create a mutant strain without an exopolysaccharide layer before the target cells would take up a transposon vector. Chemical mutagenesis is a drastic method and can result in unintended mutations; other methods of breaking up the exopolysaccharide layers should be pursued first (including possibly sonication or mild alkali treatment before electroporation). In genetic transformation research, it is also always a possibility that the bioluminescence genes and associated enzymes simply will not function or that the plasmids cannot be maintained in the host cells.

CHAPTER 1.0

INTRODUCTION

Toxic compounds in the influent of biological wastewater treatment plants can cause severe performance problems that range from poor clarifier biomass settling and elevated effluent BOD (biological oxygen demand) and ammonia levels to total plant kills. It is therefore desirable to monitor the influent to a wastewater treatment plant with the ultimate objective of taking corrective action if the toxicity rises above a level expected to degrade effluent quality.The discussion below reviews current influent monitoring practices and methods under development and presents the project team's approach.

1.1 Current Influent Toxicity Monitoring in Practice

Most municipal wastewater treatment plants (WWTPs) monitor their influent for potential toxicity by testing for specific chemical species (such as nickel, zinc, copper, chromium, etc.) and by performing a whole influent toxicity measurement using a bioassay that measures the inhibition of respiration. For the purpose of influent wastewater screening, the inhibition of bacteria is the most meaningful inhibition measurement since it is bacteria that perform the bulk of the removal of carbon and ammonia in the influent.

There are two approaches to measuring respiration inhibition. One uses a consortium of the bacteria found in activated sludge wastewater treatment plants. The other tests for respiration rate in a surrogate microorganism. In methods that use a consortium of bacteria, changes in respiration rate (normally indicated by the oxygen consumption rate) are monitored when the bacteria are exposed to influent. The bacteria that make up the consortium may be a sample of the actual bacteria from the wastewater treatment plant itself or may be a consortium isolated from typical WWTP bacteria. In either case, the test organisms are usually kept separate from the treatment plant organisms and are fed a bypass stream of influent. If the respiration rate of the test organisms drops substantially, toxicity is suspected in the influent. A highly desirable characteristic of this method is that the test organism bears a high degree of similarity to the organisms in the treatment plant. The primary disadvantages are that respirometric methods are labor intensive and that a suitable culture of bacteria must be maintained for use in the assay. Also, if the respiration rate of a sample WWTP consortium is the only variable measured, the activity of nitrifying populations will be underrepresented in terms of its effect on overall treatment performance.

In methods that use the second approach, a surrogate test for respiration inhibition in which bacteria that produce light as a function of their metabolism rate is often used to represent the actual bacteria in the treatment plant. Light is an easily measured signal compared to respiration rate, which involves measuring the time derivative of the primary signal (dissolved oxygen concentration). There are several commercially available assays, but the most common is the Microtox® Acute Assay, which is based on the bioluminescent marine bacterium *Vibrio fischeri* (also known as *Photobacterium phosphoreum*). This well-known commercially available assay uses freeze-dried cultures of the sensor organism, which are reconstituted batch-wise for the testing of individual samples.

Because the sensor organism is available in prepackaged dry form, no culture needs to be continuously maintained, and testing can occur with only a few minutes of preparation. At a municipal wastewater treatment plant frequented by the research team, the Microtox® Acute Assay typically gives EC_{50} values on the order of 50% of the original sample wastewater concentration. That is, the light produced by the test bacteria is repressed to 50% of its maximum value when the influent wastewater is diluted to half of its original concentration. At full strength, therefore, it would be expected to repress the light produced by *V. fischeri* by more than 50%.

1.2 The Biochemistry of Bioluminescence and Toxicity Monitoring

The collection of genes that produce light in *V. fischeri* and many other bioluminescent organisms are known collectively as the *lux* genes. These genes produce light more or less consistently when conditions appropriate for growth are present. The *lux* genes and their function in the production of light are shown in Figure 1-1.

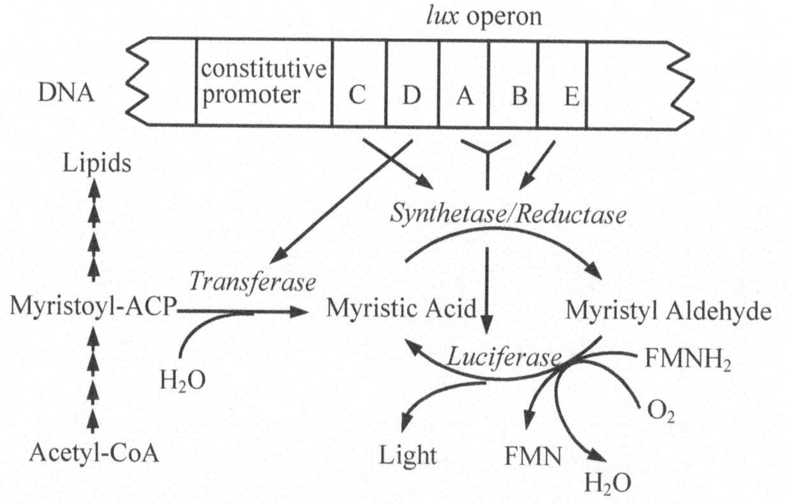

Figure 1-1. The *lux* Gene Cassette and its Associated Products.

The genes are of two types: those that bring carbon substrates from other metabolic pathways to the light-producing reaction cycle (*lux*D) and those directly involved in the cycle that produces light (*lux*A, *lux*B, *lux*C, and *lux*E). A complete review of the biochemistry of light production through the luciferase enzyme system is available in Hastings and Nealson (1977), but the basic light-producing reaction can be simplified in the following stoichiometric relationship:

$$FMNH_2 + O_2 + RCHO \xrightarrow{\text{luciferase}} 0.1hv + FMN + H_2O + RCOOH$$

which states that the long-chain aliphatic aldehyde (represented by RCHO above; in this case it is myristyl aldehyde) and the flavin $FMNH_2$ are oxidized in the presence of oxygen and the enzyme luciferase, producing light in the process. The oxidized form of the flavin FMN is regenerated (reduced) in a separate cycle by NADH in the presence of a reductase enzyme (not shown). The production of light is like most biochemical processes, a somewhat complicated process when considered in its complete detail. An additional complicating factor is the diversity of mechanisms by which toxicity influences the production of light. Toxic compounds express different mechanisms of action (MOAs) and therefore affect respiration and the production of light in different ways. Metals can bind to enzymes and reduce or increase their activity. Synthetic organic compounds can affect the membrane or compete with the natural substrates in the light producing reaction for the active sites on the enzymes. However, for the purposes of influent toxicity screening, it is not necessary to know the MOAs of the possible toxicants if one can simply correlate the response of the assay to the desired variable.

1.3 Problems with Current Whole Influent Toxicity Monitoring Methods

In the case of influent toxicity screening, the desired variable mentioned above is not a simple single variable but a composite of variables that provide a measure of the predicted effect of the influent on the performance of the wastewater treatment plant. One of the two primary problems associated with some commercially available whole influent toxicity screening assays is that they provide little guidance about what corrective action should be taken in the event of a positive toxic response. The other primary problem is that often the chosen assay does not correlate well with the desired variable or variables (i.e., the expected performance of the wastewater treatment plant). There are two types of mismatch in assay response : 1) a mismatch in the assay signal magnitude (that is, a scaling problem; for example, an assay that uniformly produces a very strong positive signal when there is little effect on the treatment system) and 2) a mismatch between toxicity profiles (that is, an assay that gives a very strong signal for some materials that produce little response from the wastewater treatment system and gives little response for some materials that produce a very strong response in the treatment system).

There are additional significant operational problems as well, such as the batch-wise and labor-intensive nature of some assays. These operational problems make the assays slow and expensive to perform as well as difficult to adapt to continuous monitoring applications. Consequently, the results from the assays are often not known until a day or more after the samples are taken so that preventative action is not possible. A continuous, online monitoring system is highly desirable so that preventative action is possible. Currently available bioluminescence-based methods, while commonly used in influent toxicity monitoring, are known to suffer from the primary problems cited above (poor correlation with plant performance and a lack of guidance for preventative action) and from the difficulty of adapting them to a continuous system.

The function of a toxicity screening system is also complicated by the nature of the variable it is trying to predict: treatment plant performance. Treatment plants perform a variety of functions designed to achieve the treatment goals of the plant. Oxygen demand, toxic metals and organic compounds, color, solids, ammonia, and microbial pathogens must all be removed to a degree consistent with the permit requirements of the facility. An upset in any one of these functions is a treatment failure and requires corrective action. By its nature, a modern biological wastewater treatment plant contains a large consortium of organisms that run the gamut of microorganism trophic levels as well as functions. In addition, organisms that primarily express a particular set of biochemical pathways can adapt to changing conditions to express a different set (for example, the effect of oxygen concentration on nitrifying bacteria). Therefore, predicting what will happen when a particular influent is fed to a plant is a difficult task for any monitoring system.

1.4 New Screening Methodologies Under Development

For the above reasons, there has been a good deal of activity in the last 10 years (and particularly in the last five years) in research directly applicable to the development of online wastewater treatment plant whole influent toxicity screening. The methodologies under development differ but usually involve either a method of monitoring the respiration rate of a surrogate organism or consortium of organisms or a method of detecting a specific microbial stress response (such as a sudden potassium efflux or the expression of genes associated with the response to a toxic environment).

1.4.1 Respiration Rate-Based Methods

A number of researchers are developing systems to monitor respiration rate more easily. Some of them have taken the approach of developing alternate means of determining the change in the concentration of oxygen in a suspended culture of bacteria (such as Chan et al., 1999), while some have concentrated on immobilizing a representative culture on the end of an appropriate electrode-based probe (such as Evans et al., 1998).

A subclass of respiration rate-based methods is bioluminescent bacteria-based methods in which the genes for bioluminescence are under the control of a constitutive promoter. As discussed above, assays based on the naturally occurring marine bacterium V. *fischeri* are of this type. However, work continues to develop new test organisms that do not suffer from the drawbacks associated with currently available methods (Lajoie et al., 2002; Gu and Choi, 2001).

1.4.2 Stress Response-Based Methods

As the name implies, these methods take advantage of the fact that bacteria respond to environmental stresses with specific actions. Bott and Love (2002) noted that certain bacteria respond to shock loads of toxic electrophilic chemicals with an efflux of potassium ions into their environment. It is known that potassium efflux is the signal for some cells to deflocculate and cause settling problems in clarification systems. These researchers are working on an upset early warning system based on this observation. Others have merged stress-response with bioluminescence genes so that cells produce light when stress-specific gene promoters are induced (Gu and Gil, 2001). For example, cells are known to respond to extremes of heat, UV light, changes in pH, DNA damaging compounds, etc., through the induction of so-called "shock genes" to produce proteins that protect the cell from these damaging elements. The *lux* genes can be placed downstream of the promoters for known shock genes so that the cells produce light if damaging conditions induce the promoters. These hybrid bioluminescent biosensors differ from those based on constitutive promoters because they produce light in response to specific conditions and not as an integrated response to toxicity.

1.5 Summary of the Current State of the Art

As discussed above, there are many methods currently used and under development for predicting the effects of toxicity on biological wastewater treatment plant performance. To date, no technology has proven effective and timely enough to allow for preventative action. All technologies suffer from their own particular drawbacks as well as those common to all methods of predicting the effects on a process as biologically diverse as an activated sludge wastewater treatment facility.

Current respirometric methods have not been widely used because they are labor intensive and measure primarily heterotrophic activity and do not effectively monitor for possible damage to nitrification. Current bioluminescence-based methods using marine bacteria tend to be more sensitive to influent toxicity than the treatment plant microorganisms themselves and are based on batch-wise assays. They also tend to correlate poorly with other conventional measures of treatment plant performance. Respirometric methods under development are improvements over current methods for measuring the time-derivative of the dissolved oxygen concentration, but they still require maintenance of a mixed culture and primarily measure the activity of heterotrophic bacteria. Stress-response methods are an interesting and promising direction but depend on a limited number of stress-response genes and provide a signal for a particular state in a single biosensor organism.

1.6 Previous Work by the Project Team

In a previous WERF-sponsored study (98-CTS-6RD), the project team used a new, genetically engineered bioreporter designated Shk1 ("Shock 1") to develop a continuous online influent toxicity monitoring system as well as a batch assay based on the same organism. The team's approach was to merge the desirable characteristics from both a native wastewater treatment bacterium and a naturally occurring bioluminescent bacterium in a single organism to be used for toxicity monitoring. The researchers found that the response of the organism to a large suite of organic toxicants as well as heavy metals compared well with literature data on activated sludge respirometry, as judged by comparing the EC_{50} values of the compounds for each of the assays.

Next, the researchers used a batch assay based on Shk1 and monitored the influent feed to a bench-scale wastewater treatment model. They periodically spiked the influent stream with toxic levels of heavy metal ions. The team compared the signal from the biosensor to the effluent COD and ammonia levels as well as the respiration rate of the activated sludge microorganisms in the aerated basin. The researchers concluded that, while the response correlated well with activated sludge respirometry, the single strain did not give an accurate prediction of the effects of toxicity on integrated measures of treatment performance such as COD removal. They also concluded that a nitrifying biosensor would have been more useful to monitor toxic effects that lead to increased effluent ammonia concentration.

1.7 The Research Approach in this Study

The search for an effective influent toxicity screening system has been likened to the search for the equivalent of a canary in a coalmine; however, it is more than that. A successful screening system needs to have a "canary" that responds appropriately to all toxic effects or several "canaries" to represent the different classes of microorganisms in the wastewater treatment plant. The approach of this study has been to utilize the best characteristics of each of the methods discussed above to produce sensor organisms that give the most detailed and appropriate response to influent toxicity.

The research team's approach was to start with bacteria indigenous to wastewater treatment plants. Then, bioluminescent bioreporters based on these bacteria were created by adding the *lux* gene cassette to the indigenous microorganisms. The objective was to develop organisms with toxic responses appropriate to the protection of activated sludge wastewater treatment organisms. In this way, the team hoped to avoid the pitfalls of other methods based on bacteria not indigenous to wastewater treatment plants. Further, they desired to add the genes to the indigenous bacteria under the control of constitutive promoters so that any toxic response affecting the metabolic rate of the bacteria would generate a response and not just responses that induced specific promoters. Biosensors from different genera were created to signal effects to as many types of organisms as possible. After the organisms were created, toxicity assays were performed to determine if each of the new biosensors provided unique information that could be included in an online multisensor influent toxicity monitoring system.

CHAPTER 2.0

CREATION OF THE BIOSENSORS

2.1 Introduction

The first objective of the project team was to use activated sludge organisms to construct the toxicity reporters using recombinant DNA techniques, rather than employing naturally occurring bioluminescent marine microorganisms or common laboratory strains. The ultimate hypothesis is that an array of indigenous activated sludge microorganisms would supply a more realistic assessment of the effects of toxicants on the activated sludge process than would individual bioluminescent reporters developed using other strains found elsewhere. The research approach includes cultivating the activated sludge isolates, purifying cloning vectors containing the bioluminescent (*lux*) genes, introducing the cloning vectors into the activated sludge strains, selecting the light-producing recombinant strains, and developing and testing a toxicity testing protocol using these strains.

2.2 Scientific Background

2.2.1 Activated Sludge Strains

The most commonly known activated sludge microorganisms are those with distinctive morphologies that are involved in sludge quality problems (filamentous bulking, foaming, and viscous sludge) and those with distinctive metabolic capabilities (nitrification). Examples of filamentous organisms include *Sphaerotilus natans* and *Haliscomenobacter hydrossis*, which can cause filamentous bulking if present in excess. *Zoogloea ramigera* forms distinctive clusters that are associated with deteriorated sludge dewatering potential, and *Gordona amarea* is involved in foaming problems. *Hyphomicrobium* spp. grow as single cells and are attached to sludge flocs via a short filament-type extension and a basal holdfast. These species are able to metabolize methanol and some other C_1 compounds, and poor settling can result from an abundance of elongated hyphae protruding from the floc. *Nitrosomonas europaea* converts ammonia to nitrite, and it and related organisms are responsible for nitrification in wastewater treatment plants. These species were chosen as candidate organisms for construction of bioreporters due to their known importance in wastewater treatment systems and their availability as pure cultures from established culture collections.

Perhaps the most important microorganisms in activated sludge are the non-descript single-celled bacteria that comprise the bulk of normal activated sludge flocs. Some of these species have been identified via the construction of 16S rDNA libraries and sequencing. However, no established representative species have been grown in pure culture or are available from culture collections.

Many of the single cell types from activated sludge that grow readily on agar plates are members of the genus *Pseudomonas. Pseudomonas* spp. are microorganisms that typically possess a considerable degree of metabolic diversity (can use a broad array of compounds as growth substrates), grow readily in rich media under common laboratory culture conditions, and are easily transformed with DNA from other species. Many of the existing bioluminescent bioreporter strains (e.g., Shk1) have been developed using members of this genus. There is some question, however, about the actual prevalence and importance of these species in the environment. Because they grow rapidly on rich media agar plates, their presence and significance in the environment may be overestimated.

Pseudomonas spp. are commonly used in developing bioluminescent bioreporters, including the previously developed toxicity bioreporter Shk1, which was isolated originally from an industrial wastewater treatment plant. Although these opportunistic species may not be representative of the most prevalent activated sludge microorganisms, they may be prevalent in episodes of dispersed growth caused by excess organic loading. These bioreporters may be useful for toxicity comparisons with other known activated sludge microorganisms.

The activated sludge environment is far different from the environment in which most common laboratory strains, such as *Eschericia coli*, are cultivated. The available organic substrate concentrations in aeration basins are typically much lower than those in rich laboratory media, such as LB, R2A and nutrient broth. Consequently, the growth rates of activated sludge microorganisms are very slow in the environment. These low growth rates are also commonly observed even if these strains are grown in rich media, suggesting that these strains have evolved to compete under low substrate conditions.

The nutritional requirements of activated sludge isolates are also not typically well established, but can usefully be augmented with vitamins. Although the culture collections from which these strains can be obtained sometimes recommend a specific medium for each strain, these media do not always support the growth of these isolates over many generations. This suggests that these media are deficient in some essential vitamin, trace metal, or amino acid.

A major concern in using nonselective media for cultivation of microorganisms is that growth by other contaminating organisms can occur more readily than in media that are somewhat selective for the desired organism. This concern is heightened when maximum growth rates of the target organisms are lower than those of many common opportunistic microorganisms, as is the case with many activated sludge microorganisms. In cases where the target organisms have distinctive morphologies, the purity of the cultures can be checked by microscopic examination. However, the typical morphologies of activated sludge microorganisms are not necessarily conserved under different cultivation conditions, adding to the uncertainty. For example, *Z. ramigera* often grows in wastewater treatment plants as zoogloeal clusters, often with finger-like extensions, with individual cells clearly visible in an exopolysaccharide matrix. In pure culture in rich media, however, these species can grow as individual dispersed cells. Under low oxygen conditions in test tubes, *S. natans* will tend to grow as a layer of filaments at the top of the liquid column where the oxygen concentration is the highest. This approach can be used to select for *S. natans*, but growth rates are much lower under these conditions than in well-aerated cultures. Because of these uncertainties, the identification of the microorganisms in the culture must be verified at some point using more advanced molecular methods such as 16S rDNA sequencing or restriction mapping.

Hyphomicrobium, *Gordona*, and *Nitrosomonas* spp. have specific metabolic capabilities that, while not unique to these genera, separate them from the most prevalent heterotrophic microorganisms. For example, *Hyphomicrobium* spp. can metabolize C_1 compounds. Members of other genera can also utilize C_1 compounds, but this capability is not generally present in microorganisms. Selective media can be employed for cultivation of *Hyphomicrobium* str. M3, *Gordona amarae*, and *Nitrosomonas europaea.* The ability of these strains to grow on media that is somewhat selective minimizes the possibility of growth of contaminants, and provides more certainty of the identity of the resulting cultures. The elongated hyphae and basal holdfast of *Hyphomicrobium* spp. can be observed under a microscope, and these traits can verify the identity of the cultured microorganism. The appearance of nitrite in the culture broth is specific to nitrifying (ammonia oxidizing) bacteria, and can provide added confidence that *Nitrosomonas* sp. is actually present in these cultures. As with the organism for which no selective media have been identified, final confirmation requires the use of 16S rDNA or other biomolecular analyses.

2.2.2 Construction of Recombinant Organisms

Genes can be transferred from one organism (donor) to another (recipient) by conjugation (commonly referred to as mating). In this naturally occurring process, the donor and recipient cells can be different species. Mating occurs between cells on stationary surfaces such as agar plates or filters, but not between cells in suspension. The genetic material is most often present on a plasmid, which is a self-replicating piece of DNA not directly associated with the chromosomal DNA. Naturally occurring plasmid DNA vectors have been modified for use in construction of recombinant microorganisms. Important elements on these vectors include an origin of replication for replication of the plasmid in the donor and/or recipient cells, a selectable antibiotic resistance gene (often ampicillin, tetracycline or kanamycin) for selection of strains containing the plasmid vector, a multi-cloning site with numerous restriction enzyme sites for insertion of the DNA of interest, and a promoter upstream of the multi-cloning site to control expression of the inserted gene of interest. Promoters can be activated by specific chemical compounds in the media (inducible promoters) or can be active continuously (constitutive promoters). The origin of replication can be narrow host range (only recognized in a limited range of species, typically *E. coli*) or broad host range (functions in a variety of species).

To conduct mating experiments, pure cultures of the donor and recipient strain can be spread-plate together on agar plates and incubated overnight. After mating, it is necessary to identify and isolate the recipient cells containing the plasmid vector. Since conjugation is a low-frequency event, it is not possible to screen all individual cells in the mating mixture to identify those containing the inserted gene of interest. Rather, approaches must be used in the next step to select against growth of both the donor cells and the recipient cells that have not actually received the plasmid DNA. One method is to streak the mating mixture on another set of agar plates containing media on which only the recipient cells containing the plasmid vector will grow. To select against the donor strain, an antibiotic can be used that inhibits the donor cells, but to which the recipient cells are not sensitive. An alternative is to use a minimal media with a growth substrate that supports the growth of the recipient cells, but not the donor cells. To select against recipient cells lacking the plasmid being transferred, some trait conferred by the plasmid must be used to select for strains containing the plasmid. This can be an antibiotic resistance gene for an antibiotic to which the recipient strain is normally sensitive or a catabolic gene that allows the recipient cells to grow on a substrate that they cannot otherwise metabolize.

An alternative to conjugation is electroporation, which employs short-duration pulses of electric current to create pores in the cell membrane through which exogenous DNA can enter. After the electric pulse is discontinued, the pores close in approximately 1 to 30 minutes. After a recovery period in quiescent culture broth, the cells can be plated on selective media. For these experiments, purified plasmid DNA is mixed with the cells prior to electroporation.

The advantage of electroporation over mating is that no donor cells are present, so no counter-selection method is needed to eliminate these cells in the final selective media. All that is required are media for selection of strains successfully transformed with the exogenous DNA. This technique is most effective for smaller size plasmids, which can be easily extracted from the donor cells (typically *E. coli*) and can enter the recipient cells as circular plasmids. Electroporation can sometimes be made more efficient by linearizing the vector DNA by hydrolysis with restriction enzymes. The process can also be optimized for cells from different species by adjusting the DNA concentration, capacitance, and voltage applied to the electroporation cuvette.

In addition to plasmids, another common cloning and mating vector is the transposon. Unlike plasmids, which are autonomous DNA sequences, transposons or transposable elements can integrate into the chromosome of the recipient host. Because they become part of the host chromosome, transposons are often more stably maintained in the cell than are plasmids and are therefore sometimes preferable. The integration of the transposon into the host DNA is catalyzed by enzymes called transposases, which recognize inverted terminal repeats bracketing the transposon.

Naturally occurring transposons have been modified for use in the construction of recombinant microorganisms. Transposons, including nucleotide sequences for the terminal repeats and transposase, are usually cloned into plasmids with narrow host range *E. coli* origins of replication to maintain and prepare DNA for transformation of other species. Within the terminal repeats are an antibiotic resistance gene and a multi-cloning site for insertion of the gene of interest. The cloning site may be downstream of a promoter or the construct may be promotorless. The plasmid containing the transposon is transformed into the target species by mating, as described above. The transposase catalyzes the insertion of the sequences between the terminal repeats, including the cloned gene of interest and the antibiotic resistance gene, into the chromosome of the target species. Since the origin of replication is narrow host range, the plasmid itself is not replicated during cell division, and is lost from all daughter cells. Only those cells in which chromosomal insertion occurs will contain the cloned gene and antibiotic resistance gene. The transposase is typically located outside the inverted repeats, so that once the plasmid is lost from the cell, no further transposition of the cloned gene occurs.

Integration of transposons typically occurs at random locations in the host chromosome. If the transposon construct contains a promoter, then most of the recombinant cells will produce similar amounts of the enzyme encoded by the cloned gene. If the transposon construct is promotorless, then the timing and rate of gene expression will be dependent on the endogenous promoter upstream of the insertion location, and different recombinant cells will produce different amounts of enzyme, depending on the insertion location.

2.2.3 Vectors and Genes for Light Production

Several vectors containing the *lux* operon have been developed. Three of these vectors that the project team have found to be particularly suitable for insertion of the *lux* genes in environmental isolates are the transposon vectors Tn4431 (see Shaw et al., 1987) and mini-Tn5 *luxAB* and the plasmid vector pUTK2 (Figures 2-1 and 2-2).

Tn4431 contains the promoterless *lux* operon and a tetracycline resistance gene within inverted terminal repeats on a narrow host range *E. coli* plasmid. This plasmid also contains an ampicillin resistance gene (outside of the inverted repeats). Light levels are dependent on the insertion location in the recipient chromosome. Vector mini-Tn5 *luxAB* also lacks a promoter upstream of the *lux* genes and contains the tetracycline resistance gene. However, this vector contains only the *luxAB* genes, so light production is dependent on adding the substrate aldehyde to the culture broth. Plasmid pUTK2 was developed by inserting the transposon Tn4431, containing the *lux* genes, into a naturally occurring plasmid. The *lux* genes are located downstream of a constitutive plasmid maintenance promoter, and recombinant strains developed using this plasmid typically produce high light levels. The vector lacks transposase, so there is no integration into the host chromosome. This plasmid is often stably maintained in recipient cells. The previously developed Shk1 toxicity bioreporter was developed using this plasmid. All of these vectors are routinely maintained and prepared from *E. coli* cultures.

The genes encoding bioluminescence have been previously inserted into the nitrifying (ammonia oxidizing) bacteria *Nitrosomonas europaea* (Iizumi et al., 1998) using plasmid pHLUX20 (Figure 2-3). The *luxAB* genes are located on a plasmid downstream of a hydroxylamine oxidoreductase (HAO) promoter.

Since this recombinant strain lacks the other genes in the *lux* operon, light production requires the addition of aldehyde to the culture broth. This plasmid carries a gene that confers kanamycin resistance. Unfortunately, the recombinant *N. europaea* strain that produces bioluminescence has not been maintained and is no longer available.

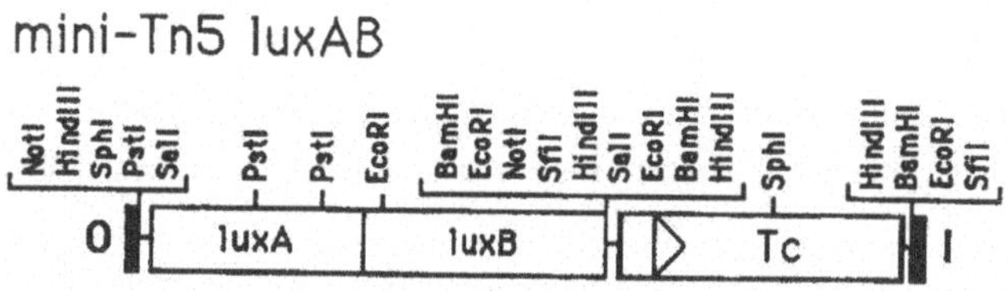

Figure 2-1. Mini-Tn5 *luxAB* Transposon. This Transposon is Carried on a R6K-based Suicide Delivery Plasmid (not shown). (Reprinted with permission from American Society for Microbiology)

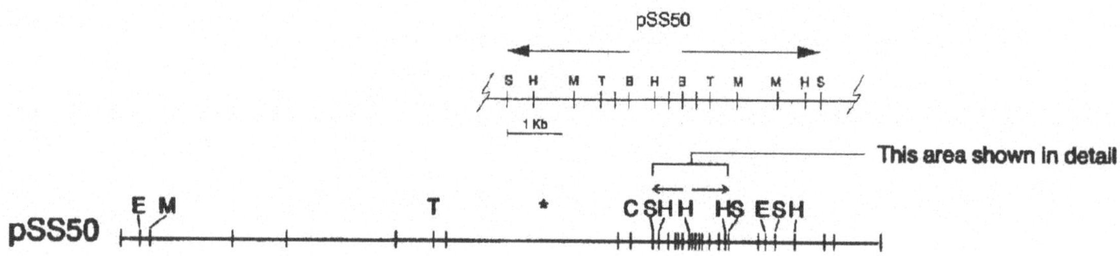

Figure 2-2. Plasmid pSS50 into Which Was Inserted the *lux* Operon (transposon Tn4431) to Create Plasmid pUTK2. The Insertion Location of the *lux* Operon is Indicated with a "*".
(Reprinted with permission from American Society for Microbiology)

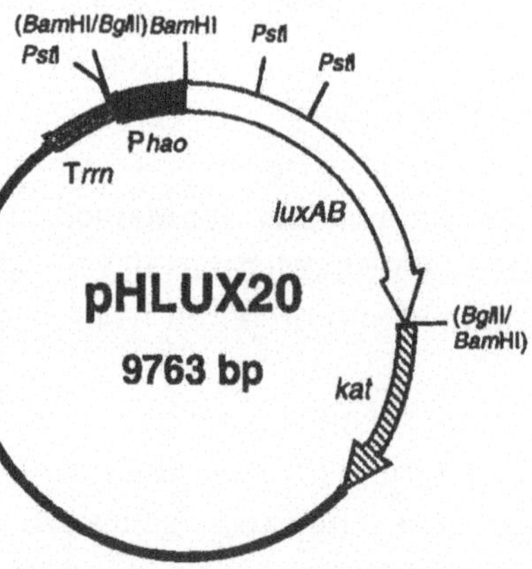

Figure 2-3. Physical Map of Plasmid pHLUX20.
(Reprinted with permission from American Society for Microbiology)

2.3 General Methods

2.3.1 Sources of Activated Sludge Strains

The researchers obtained the following strains from the American Type Culture Collection; *Zoogloea ramigera* (type strain; ATCC 19544), *Sphaerotilus natans* (type strain; ATCC 13338), *Haliscomenobacter hydrossis* (type strain; ATCC 27775), *Gordona amarae* (ATCC 27808), *Zoogloea ramigera* (ATCC 19623), *Nitrosomonas europaea* (ATCC 19178), *Duganella zoogloeoides* (ATCC 25935), *Pseudomonas putida* (ATCC 12633), and *Zoogloea* sp. (ATCC 19324). *Hyphomicrobium* strain M3 (ATCC 202122) was previously isolated from the Eastman Chemical Company wastewater treatment plant in Kingsport, TN (Layton et al., 2000), and was obtained directly from Arthur Meyers at Eastman Chemical Company. For the isolation of *Pseudomonas* spp., the researchers obtained activated sludge from the aeration basin of the Syracuse Metropolitan Area Wastewater Treatment Plant (METRO) in Syracuse, New York. The project team mixed the activated sludge in a blender to disperse the cells in the activated sludge flocs. Serial dilutions of the sludge were prepared in phosphate buffered saline (PBS), and dilutions were plated on *Pseudomonas* isolation agar (PIA, Appendix A).

2.3.2 Sources of *lux* Genes and Vectors

The project team obtained:

- *E. coli* containing pUCD623 (*lux*-transposon Tn4431) from Dr. Clarence I. Kado, University of California, Davis, Department of Plant Pathology, One Shields Avenue, 354 Hutchison Hall, Davis, CA 95616.
- The mini-Tn5 transposon derivatives CC118 λpir/pUT-mini-Tn5 Km1 and pUT-mini-Tn5 *lux*AB from Dr. Carlos A. Guzman, Leader Vaccine Research Group, Department of Microbial Pathogenesis and Vaccine Research, Division of Microbiology, GBF-German Research Centre for Biotechnology, Mascheroder Weg 1, D-38124 Braunschweig, Germany.
- *E. coli* containing plasmid pUTK2 from the University of Tennessee culture collection.
- *E. coli* containing the plasmid construct pHLUX20 from Dr. Iizumi at Kurita Water Industries Ltd in Japan. These *E. coli* strains were all grown on Luria-Bertani (LB) medium and stored at $-80^{o}C$.

2.3.3 Cultivation of Activated Sludge Strains

In order to prepare individual activated sludge microorganisms for transformation with the *lux* operon, suitable culture media must be identified. The same media amended with selective antibiotics could also serve for routine cultivation of the organisms for toxicity testing of influent, activated sludge, and effluent samples. The purpose of these experiments was to find suitable growth conditions for the cultivation of the activated sludge strains chosen for use in this study.

Freeze-dried cultures of strains obtained from ATCC were resuspended by the project team in PBS and plated on several different agar media plates, including media recommended by ATCC for each strain as well as other media described in the scientific literature where available (Appendix A). The researchers then transferred cells from the colonies to liquid media of the same composition to cultivate for storage (freezing in glycerol at –80°C) or use in transformation experiments.

2.3.4 Screening for Antibiotic Sensitivity

As discussed earlier, to be successful mating and electroporation experiments must employ a method to select against the recipient cells that do not receive the target DNA (plasmid or transposon). Also required is a method for selecting recipient strains expressing some factor (e.g., antibiotic resistance) encoded by a gene on the DNA being transferred. All of the *lux*-containing vectors described above (mini-Tn5 *luxAB*, Tn4431, and pUTK2) contain the tetracycline resistance gene as the selectable marker, so initially the recipients must be sensitive to tetracycline prior to transformation.

To determine antibiotic sensitivities, the project team prepared agar plates containing different concentrations of tetracycline. They spread the activated sludge strains on these plates, and noted the presence or absence of growth at room temperature.

2.3.5 Mating

One approach used to construct the bioluminescent-activated sludge bioreporters was mating (conjugation). *E. coli* DH5α Tn5 *luxAB*, *E. coli* DH5α Tn4431, and *E. col*i S17-λpir with pUTK2 were grown in LB medium with 10 mg/L tetracycline and cultured at 37°C overnight. Recipient activated sludge strains were cultivated on the medium most effective for growth of each strain at room temperature. After growth of the cultures, *E. coli* containing the vector and the activated sludge recipient were spread-plated together onto nutrient broth agar plates. The plates were incubated at room temperature for two days. The cells were collected by scraping with an inoculation loop, and resuspended in PBS. The cell suspension was then plated onto the final selective agar medium (selects against *E. coli*) with 25 mg/L tetracycline (selects against recipients not receiving the vector), and incubated at room temperature for two weeks. Individual colonies on the plates were picked and streaked as lines (50 lines/plate) on the same medium with tetracycline, and incubated at room temperature. Strains producing visible light were detected in a dark room and restreaked. Morphologies of the cells in resulting colonies were checked microscopically.

Mating was the chosen method for introducing the *lux* genes into *G. amarae, N. europaea, H.* str. M3 and the three *Pseudomonas* spp. (recipient) because these strains grow on media that do not support the growth of *E. coli* (donor) (Appendix A). *G. amarae, H.* str. M3 and the three *Pseudomonas* spp. were mated with *E. coli* strains containing the mini-Tn5 transposon derivatives CC118 λpir/pUT-mini-Tn5 Km1 and pUT-mini-Tn5 *lux*AB, pUTK2 and pUCD623 (*lux*-transposon Tn4431). *N. europaea* was mated with an *E. coli* strain containing pHLUX20. The final selection media for *G. amarae, N. europaea, H.* str. M3 and the three *Pseudomonas* spp. were CD, P, IMH, and PIA respectively (Appendix A), and contained either 10 or 25 mg/L tetracycline. The research team did not use matings in transformation experiments with the other activated sludge strains because they could not identify suitable selective media.

The highest light producing strains were identified and grown at room temperature in 100 ml of the same medium in shake flasks. Every six hours, the optical density in each culture was measured by spectrophotometer, and the bioluminescence was measured by luminometer.

2.3.6 Electroporation

The project team chose electroporation to try to introduce the *lux* genes [pUCD623 (*lux*-transposon Tn4431)] into *Zoogloea ramigera* (type strain; ATCC 19544), *Sphaerotilus natans* (type strain; ATCC 13338), *Gordona amarae* (ATCC 27808), *Zoogloea ramigera* (ATCC 19623), *Duganella zoogloeoides* (ATCC 25935), *Pseudomonas putida* (ATCC 12633), and *Zoogloea* sp. (ATCC 19324). Electroporation was also used to introduce pHLUX20 into *Nitrosomonas europaea* (ATCC 19178).

For electroporation experiments, *E. coli* strains containing the cloning vectors with the *lux* genes were grown on LB medium with tetracycline. The cells were then centrifuged and lysed to release the plasmids. The plasmids were purified using the S.N.A.P. midi-prep procedure from Invitrogen, Inc. (Carlsbad, CA). The recipient activated sludge strains were grown on the appropriate medium, and then centrifuged and washed with cold (4°C) distilled water and a 10% glycerol/water solution.

The electroporator voltage was set at 2.5 kV, minimum resistance at 200 Ω and maximum resistance at $\infty\ \Omega$. The field strength was 1.25 kv/mm gap. Ten µL of circular or linearized plamid DNA (1.6 µg of DNA) was added into tubes containing 100 µL electrocompetent cells. The mixture of cells and DNA was transferred to a prechilled 2 mm gap cuvette. The cuvette was inserted into the cuvette holder and placed into the electroporator. The electrical pulse was applied, and then the cells and DNA were mixed with 1 mL of growth medium and incubated at 30°C for 2-12 hours. After incubation, all the cells were plated on solid medium (with 25 mg/L tetracycline for selection). When colonies appeared on the experiment plate (with DNA added), they were restreaked on another agar plate with 25 mg/L tetracycline and incubated at 30°C. After seven days, individual colonies from each strain were inoculated into 1 mL of liquid medium, and the bioluminescence was measured at 60, 120, and 180 minutes. The highest light-producing isolates were regrown in liquid medium, and cell density and bioluminescence were measured with time. Alternative electroporation voltages, pretreatment of the recipient strains by vortexing with glass beads to remove exopolysaccharide coatings or sheaths, linearization of the vectors by restriction digests, different cell densities, and different antibiotic concentrations were attempted for strains that were found to be difficult to transform.

2.3.7 Preliminary Toxicity Screening at Syracuse University

Before transferring transformant strains to team members at the University of Tennessee (UT), the team members at Syracuse University performed their own preliminary screening on most of the strains (all three of the *Hyphomicrobium* strains and two of the *Pseudomonas* strains) to determine if they had suitable characteristics (sufficient light production and predictable dose-response to a model toxicant) to warrant further testing by team members at UT. The Syracuse University team members used a toxicity screening method designed to provide each of the bioreporters at the same cell density and to use the same dilution medium (either synthetic wastewater or actual wastewater) for the toxicants and bioreporters.

Hyphomicrobium Tn4431 H1, H2, and H3 were grown in 100 mL IMH medium with 10 mg/L tetracycline. Strains PM2 pUTK2 and PM5 pUTK2 were grown in 100 mL NB medium with 10 mg/L tetracycline. Every six hours, the optical density in each culture was measured by spectrophotometer, and the bioluminescence was measured by luminometer. When the OD_{600nm} reached 0.9-1.0, 25 mL of cell culture were transferred into another flask and stored at 4°C for four days. The 25 mL samples of refrigerated culture were centrifuged at 14,000*g* for 10 minutes, and then the cell pellets were resuspended in 4°C 0.1M KCl to an OD_{600nm}=1. The resuspended cultures were maintained at 4°C for two days. One mL of these refrigerated cells from each flask was transferred to a luminometer tube and placed in a room-temperature water bath. The bioluminescence from each tube was measured every 20 minutes for 120 minutes to observe light production as the cells warmed to room temperature.

Serial dilutions of zinc chloride were prepared in 0.1M KCl to achieve concentrations of 250, 125, 62.5, 31.25, 15.6, 7.8, 3.9, 1.8, and 0.9 mg/L Zn^{2+}. Eight hundred µL of either synthetic wastewater (see below for composition) or actual wastewater and 100 µL of each $ZnCl_2$ dilution were added to luminometer tubes, followed by 100µL of cell culture (after 20 minutes of incubation in a room temperature water bath). The bioluminescence was measured seven minutes after adding the cells with a tube luminometer (MiniLumat LB 9506, E,G&G Berthold, Bad Wildbad, Germany). Final Zn^{2+} concentrations tested were 25, 12.5, 6.25, 3.13, 1.56, 0.78, 0.39, 0.18, and 0.09 mg/L Zn^{2+}.

The synthetic wastewater (SWW) was a slightly modified version of that found in Kaiser et al. (1997) and contained the following (in mg/L): urea, 47; nutrient broth, 75; soluble starch, 7.5; dried milk, 75; CH_3COONa, 25; $Na(HCO_3)$, 12.5; K_2HPO_4, 3.75; $FeSO_4 \cdot 7H_2O$, 0.63, Na_2EDTA, 0.074; $CuSO_4 \cdot 5H_2O$, 0.159; $CoCl_2 \cdot 6H_2O$, 0.018; $ZnCl_2$, 0.062; $MnSO_4.H_2O$, 0.02; $MoNa_2O_4 \cdot 2H_2O$, 0.005; and $NiSO_4 \cdot 6H_2O$, 0.1. The actual wastewater was collected from the influent to the METRO wastewater treatment plant in Syracuse, NY, and is designated MWW.

2.4 Discussion of Results

2.4.1 Cultivation of Activated Sludge Strains

The project team was able to culture all of the activated sludge strains obtained, with the exception of *H. hydrossis*. Results of growth tests of some of the different activated sludge strains on different nonselective agar media are indicated in Table 2-1.

Table 2-1 Growth of Activated Sludge Strains on Nonselective Agar Media.

Strains/media	Zoo	13338	1790	1858	R2A	NB	T	AS
Z. ramigera	+	-	-	+	+	+	++	-
S. natans	+	-	-	+	+	+	++	
H. hydrossis	-	-	-	-	nt	nt	-	-
H. str. M3	-	-	nt	+/-	+	-	nt	
G. amarae	nt	nt	nt	nt	nt	nt	++	

"+"= moderate growth on specified agar medium
"-"= no growth
"+/-"=weak, slow growth
"++"= fast growth
"nt"= not tested due to known incompatibility

Z. ramigera and *S. natans* grew on Zoo, 1858, NB, and T media. Inoculation of individual colonies into the same liquid medium resulted in growth, but resulting cell densities were often fairly low (as indicated by culture turbidity), and the growth of cultures tended to decrease with each subsequent transfer to new media. The fastest growth was on T medium at 30°C (Figure 2-4), and thus far, these cultures appear to be most stable over time. *G. amarae* also grew on T medium. *H. hydrossis* did not grow on any of the media tested.

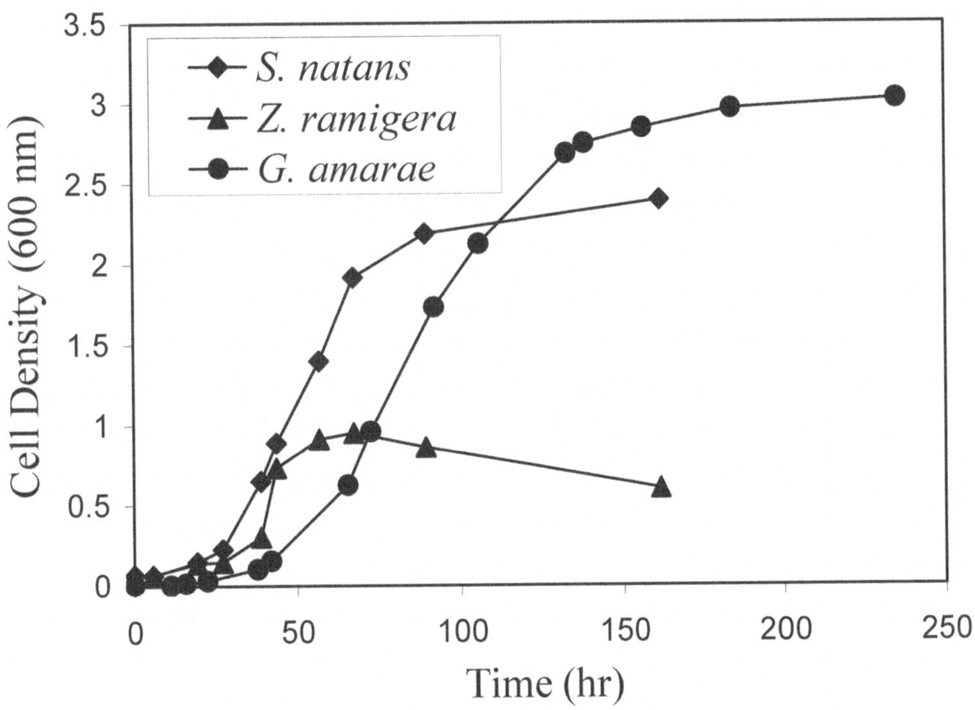

Figure 2-4. Growth of *S. natans, G. amarae,* and *Z. ramigera* on T Medium in Shake Flask Cultures.

In experiments with selective media, *G. amarae* grew on CD medium and *H.* str. M3 grew on IMH medium. *N. europaea* grew on P medium. Results of nitrite measurements in liquid cultures of *N. europaea* indicated growth. *Zoogloea ramigera* ATCC 19623 grew on nutrient broth; *Duganella zoogloeoides* ATCC 25935 grew on trypticase soy medium (ATCC medium 18), *Pseudomonas putida* ATCC 12633 grew on nutrient broth, and *Zoogloea* sp. ATCC 19324 grew on zoogloea medium (ATCC medium 189), all at 30°C on a rotary shaker. Of particular interest for subsequent mating experiments was the observation that *G. amarae, N. europaea,* and *H.* str. M3 (recipients) grew on media that do not support the growth of *E. coli* (donor). Similarly, the three *Pseudomonas* spp. isolated from the METRO activated sludge plant grew on PIA, which contains an antibiotic that is lethal to *E. coli*.

Although cultivation of candidate activated sludge organisms was not always convenient, due to low growth rates, cultures of several desirable activated sludge organisms could be reliably produced for subsequent gene transfer experiments to produce bioluminescent bioreporters. The identity of the cultures of *S. natans* could be verified by its growth form [It grows as white layer on the top of stationary (no mixing) cultures] and microscopically by its filamentous morphology. *Hyphomicrobium* could be identified microscopically by its characteristic hold-fast. *Nitrosomonas europeae* growth could be verified by the production of nitrite. Of all the cultures, it was most difficult to verify whether the Zoogloeal types contained the expected species.. Although they grow as exopolysaccharide-containing clusters in wastewater treatment systems, they grow as dispersed cells in rich culture media, and have no other distinct morphological traits. So there was always some uncertainty as to the true identity of these species, other than the confidence provided by good sterile technique. This is not a serious problem because, if a useful bioluminescent reporter were developed from one of these strains, the species could be definitively identified using bimolecular methods.

2.4.2 Screening for Antibiotic Sensitivity

The project team found all of the activated sludge strains to be tetracycline sensitive. Since all of the cloning vectors used in this study contained the tetracycline resistance gene as a selectable marker, it is useful that the activated sludge strains are tetracycline sensitive before transformation with the vector. This allows easy detection of the bacterial cells that received and can express the tetracycline genes on the vector (also containing the *lux* genes), since these are the only cells that will grow on agar plates containing tetracycline.

2.4.3 Mating

The project team found many large colonies on the plates resulting from the mating of *Hyphomicrobium* and *E. coli* Tn4431. After regrowing these colonies, the team checked them visually in the dark room for light production. The three brightest colonies were restreaked on IMH with 25 mg/L tetracycline agar plates. These three isolates were then grown on IMH medium with 25 mg/L tetracycline and stored at –80°C in 15% glycerol.

Figure 2-5 shows the results of growth and bioluminescence measurements in IMH liquid medium in shake flasks. Strains H1 and H3 entered stationary phase 145 hours after inoculation. The light production of H1 decreased when the culture entered stationary phase. Light production in strains H2 and H3 decreased while the cultures were still in the exponential growth phase.

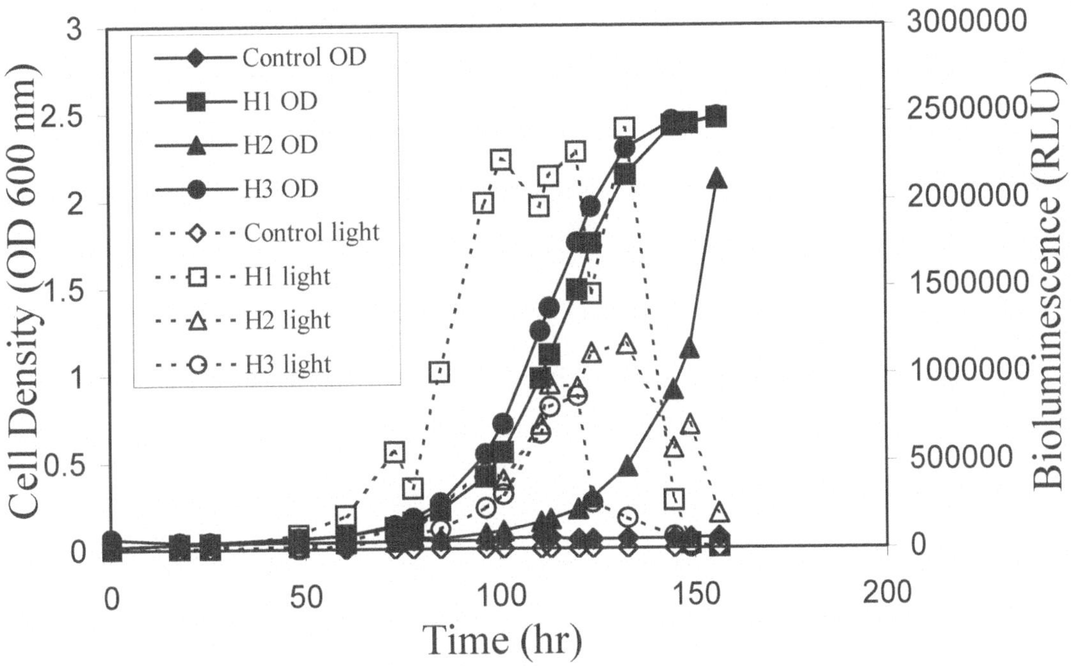

Figure 2-5. Growth and Light Production by *Hyphomicrobium* str. M3 Bioreporters Containing Transposon Tn4431.

Matings between *Hyphomicrobium* and *E. coli* pUTK2 were not successful, suggesting that the plasmid pUTK2 cannot be maintained (replicated) in *Hyphomicrobium.* Different levels of light are produced by different *Hyphomicrobium* isolates with Tn4431 because the transposon integrates into different locations in the host chromosome. The transposon must integrate into the chromosome of the host cells to replicate, as the plasmid carrier is a narrow host range plasmid. The different locations of integration result in different levels of transcription of the downstream genes (*lux* operon), depending on the particular upstream promoter responsible for transcription. The nature of the promoter could also affect the timing of transcription (e.g., growth vs. stationary phase), and presumably transcription could be affected by specific inducers of the promoter in the medium or wastewater. Nonetheless, this approach can be very effective in finding strong promoters that result in high light output in strains for which no suitable plasmids and promoters are available.

The cells grew to an OD_{600nm} of 2.5 in approximately 140 hours. H2 had a longer lag phase than H1 or H3, but once it began growing, had a growth rate similar to the other two isolates. Light production by H2 and H3 was lower than H1.

The results of the mating experiments with the three *Pseudomonas* spp. are shown in Table 2-2. No colonies were seen from the control mating experiments. The investigators successfully inserted bioluminescence genes into PM 2,5, and 6. After restreaking individual colonies on PIA agar plates with 25 mg/L tetracycline, bioluminescence was observed in a dark room. The brightest colonies (PM2 pUTK2, PM5 pUTK2, PM6 pUTK2, and PM6 Tn4431) were grown on NB medium with 25 mg/L tetracycline and stored in glycerol at –80°C.

Table 2-2. Colonies Present on PIA Agar Plates after Mating and Incubation.

	none	PM2	PM4	PM5	PM6
None	-	-	-	-	-
Tn4431	-	-	-	-	+
Tn5 *luxAB*	-	-	-	-	+
pUTK2	-	+	-	+	+

"+"= colonies established
"-"= no colonies seen

Figure 2-6 graphs the relationships between growth and light production for two of the bioluminescent *Pseudomonas* strains.

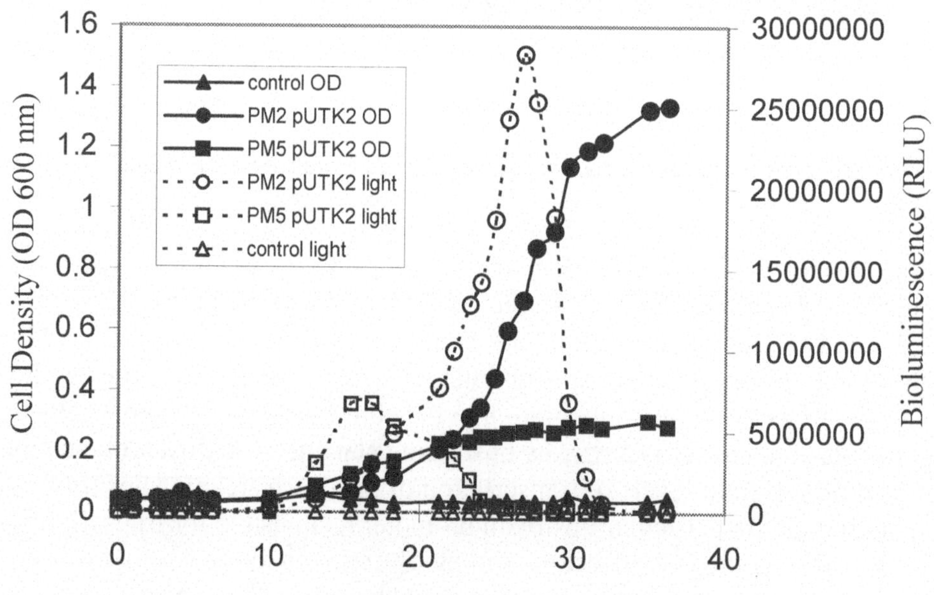

Figure 2-6. Relationship Between Growth and Bioluminescence for the Two Bioluminescent *Pseudomonas* Strains PM2 pUTK2 and PM5 pUTK2.

Pseudomonas strain PM2 pUTK2 grew to a much higher density and produced more light than PM5 pUTK2. The bioluminescence maximum occurred at an OD_{600nm} of approximately 0.8-0.9 after 27 hours incubation. At higher cell densities, bioluminescence decreased.

The project team successfully inserted plasmid pUTK2 into three *Pseudomonas* strains and observed isolates with readily detectable bioluminescence. Since the *lux* genes are placed downstream of a constitutive promoter in pUTK2, most recombinant strains for each species are expected to produce similar light levels. The transposons containing the *lux* genes were successfully inserted only into strain PM6. These transposons do not contain their own promoters, so the amount of light produced depends on the location of insertion in the host chromosome. The project team detected one recombinant PM6 strain that produced considerably more light than the other isolates, and this strain was stored at –80°C for possible future study. The identification of these species as *Pseudomonas* spp. is based on their ability to grow on PIA. More precise identification will require sequencing of the 16S rDNA.

Matings between *E. coli* and *G. amarae* and *N. europaea* did not result in the growth of any tetracycline-resistant colonies on the final selective media. Either the vectors were not transferred efficiently between the two different species or the tetracycline resistance gene was not expressed properly in these hosts.

2.4.4 Electroporation

The project team found that electroporation was not successful for introducing the *lux* vectors into any of the activated sludge strains, with the exception of the *Pseudomonas* spp. isolated from the METRO wastewater treatment plant in Syracuse, New York. Control electroporation experiments with *E. coli* and the same vectors were also successful.

It is unclear why it was not possible to transform *Z. ramigera* (and similar species), *S. natans,* and *G. amarae* cells with the vector DNA containing the *lux* genes. *G. amarae* is Gram positive, and the different cell wall may require higher voltages. However, a range of voltages was attempted without any positive results. It is possible that the sheath on the *S. natans* filaments and the exopoysaccharides produced by *Z. ramigera* inhibit the uptake of exogenous DNA. The research team tried to disrupt the exopolysaccharide layer and the sheath by blending prior to electroporation, but still they obtained no transformants. They also tried unsuccessfully to increase the efficiency of uptake by using linearized DNA in electroporation experiments. Electroporation experiments were conducted primarily with [pUCD623 (*lux*-transposon Tn4431)], since this vector both integrates in the chromosome via a transposon and transposase and also contains all the *lux* genes required to make light. However, this is a relatively large vector compared to the mini-Tn5 transposons, and this size can decrease electroporation efficiency.

2.4.5 Results of Preliminary Toxicity Screening at Syracuse University

The project team performed preliminary toxicity screening using the three *Hyphomicrobium* bioluminescent bioreporters (H1, H2, and H3) and the two *Pseudomonas* strains PM2 pUTK2 and PM5 pUTK2 described above. The

Hyphomicrobium strains were grown in IMH medium with 10 mg/L tetracycline, centrifuged, resuspended in 0.1M KCl, and refrigerated. The *Pseudomonas* strains were grown on nutrient broth with tetracycline and treated in the same manner.

The light production by all three refrigerated *Hyphomicrobium* strains increased after incubation at room temperature (Figure 2-7). No further increases were seen after 20 minutes of incubation. The toxicity testing results with the three *Hyphomicrobium* strains were similar (see Figure 2-8 for an example of the dose response curves for a *Hyphomicrobium* strain used to obtain the EC_{50} values), and they are summarized in Table 2-3. The results of Zn^{2+} toxicity testing for the two *Pseudomonas* strains were also in the same range (see Figure 2-9 for an example of the dose response curves for a *Pseudomonas* strain used to obtain the EC_{50} values), and they are summarized in Table 2-4.

Table 2-3. Zn^{2+} EC_{50} of *Hyphomicrobium* str. H1, H2, and H3 in Synthetic (SWW) and METRO (MWW) Influent Wastewater.

Strain	Zn^{2+} EC_{50} (mg/L)	
	SWW	MWW
H1	0.52	1.93
H2	0.68	1.34
H3	0.43	1.32

Table 2-4. Zn^{2+} EC_{50} of *Pseudomonas* str. PM2 pUTK2 and PM5 pUTK2 in Synthetic (SWW) and METRO (MWW) Influent Wastewater.

Strain	Zn^{2+} EC_{50} (mg/L)	
	SWW	MWW
PM2 pUTK2	0.494	2.08
PM5 pUTK2	0.18	2.32

Cell culture samples for these screening assays were collected during the exponential growth phase when light production is highest. Both the *Hyphomicrobium* and *Pseudomonas* strains produced measurable light levels after refrigeration. Based on the light recovery curves after refrigeration (see for example Figure 2-7), the investigators chose 20 minutes as the standard room temperature incubation time for the refrigerated cells before addition to the toxicity assays. The Zn^{2+} EC_{50} in synthetic wastewater for all strains was significantly lower than in METRO wastewater. This indicates that natural wastewater (MWW) is probably better able than synthetic wastewater to bind Zn^{2+}.

The five strains represented in Tables 2-3 and 2-4, as well as the additional *Pseudomonas* biosensor (designated PM6, created by chromosomal insertion of the *lux* genes) were transferred to UT for further testing. Preliminary screening at UT to identify the best candidates for detailed testing against a larger suite of toxicants is discussed in Appendix B. Chapter 3.0 discusses the more detailed testing of the best candidate strain identified in Appendix B and its comparison to other assays. In the assays described in Appendix B and Chapter 3.0, the investigators used distilled water as the dilution background liquid for the cells and toxicants. In contrast, the preliminary studies conducted at Syracuse University used synthetic and natural wastewater.

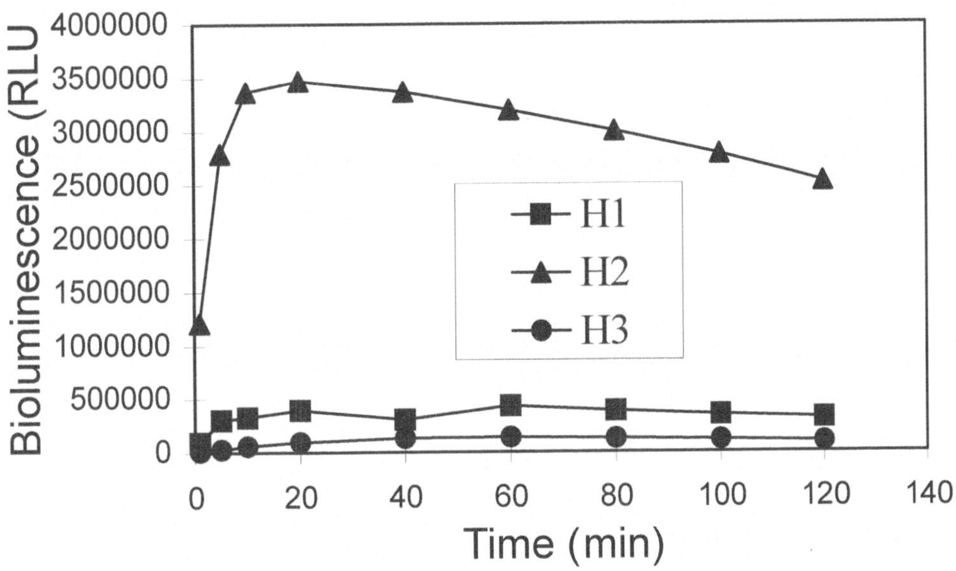

Figure 2-7. Bioluminescence of *Hyphomicrobium* str. M3 Bioreporters H1, H2, and H3 at Room Temperature After Refrigeration.

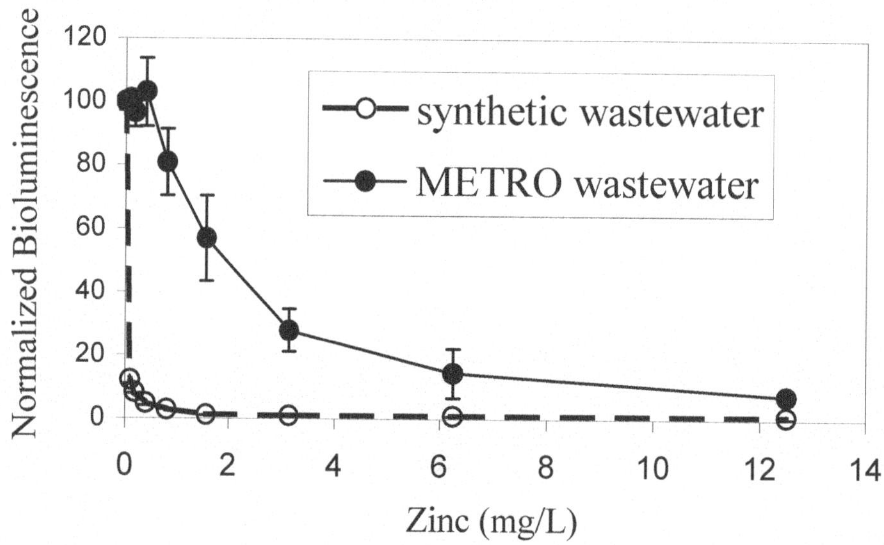

Figure 2-8. Effect of Zn^{2+} on Bioluminescence of the *Hyphomicrobium* str. M3 Bioreporter H1.

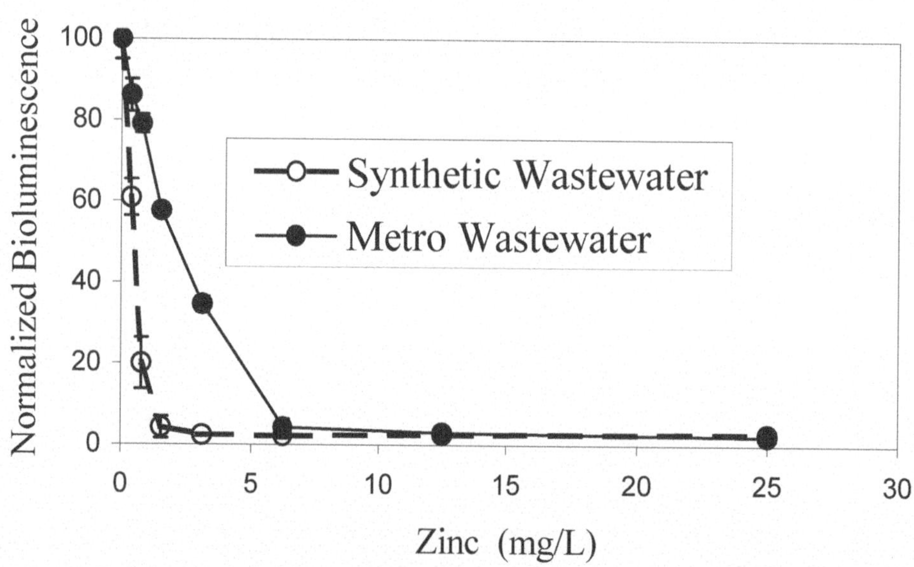

Figure 2-9. Effect of Zn^{2+} on Bioluminescence from *Pseudomonas* str. PM2 pUTK2.

Distilled water was used to minimize the effects of metals chelation by the solids in wastewater and thereby to allow these results to be compared to literature data on a commercially available assay using the naturally occurring bioluminescent bacterium *Vibrio fischeri.*

CHAPTER 3.0

TOXICITY TESTING OF THE BIOSENSORS

3.1 Introduction

As discussed in Chapter 2.0, members of the project team at Syracuse University created several bioluminescent sensor bacteria. The project team transferred the most promising of these sensor bacteria to the members of the project team at the University of Tennessee (UT); these included three *Pseudomonas* strains denoted PM2, PM5, and PM6, and three *Hyphomicrobium* strains denoted H1, H2, and H3. The project team at UT generated toxicity data for these strains using metals and organic compounds that frequently appear in wastewater. Toxicity data were collected 1) to determine which bioreporters would be good candidates for toxicity monitoring and 2) determine which collection of the new biosensors would be appropriate to include in an array of biosensors with previously constructed Shk1 (a *Pseudomonad*).

3.2 Development of the Final Toxicity Testing Protocol

All strains were grown in their respective preferred growth media to an optical density measured at 600 nm of 1. In order to provide a consistent background medium, the cells of each strain were then separated from their growth media, resuspended in potassium chloride, and stored separately at 4°C. In an earlier study (Lajoie et al., 2002), the researchers had found that storage at 4°C of the bioreporter Shk1 was a convenient way to keep cells viable. This allowed the researchers to perform toxicity assays daily without reestablishing a fresh culture for each day of experiments. Cells stored in this way could be used for up to a month before it was necessary to reculture a fresh stock.

The researchers conducted initial screening experiments (described in Appendix B) and found that all of the newly created sensor organisms produced less luminescence than Shk1 in the batch-wise assay protocol. PM6 produced a light level approximately one tenth of that produced by Shk1 under identical conditions, and PM2, PM5, H1, H2, and H3 yielded light levels even lower. A low light level is not desirable if it decreases the signal-to-noise ratio to an unsatisfactory level (which it did in this case; the signal-to-noise ratio was close to 1 for some samples). In addition, the low level of light made it necessary to use a high concentration of cells with each test sample, an undesirable condition, since it increases the absorption of toxicants on the biosensor. This increase often leads to a reduction in the observed toxicity, as shown by Franklin et al. (2002). Those earlier researchers studied the effect of cell density in algal bioassays on toxicant bioavailability and toxicity. For these reasons, the project team found it necessary to eliminate the strains H1, H2, H3, PM2, and PM5 from further consideration as viable toxicity monitoring biosensors. From this point on, they focused on obtaining toxicity data for Shk1 and PM6 in batch-wise assays. They were therefore able to decrease the cell concentration in the test solutions for the final toxicity testing protocol (described in 3.3 Methods) and still achieve a satisfactory light level.

3.3 Methods

As Ribo (1997) pointed out, one of the first steps in validating a new assay is to compare its results with the results of existing assays. Since activated sludge respiration or growth inhibition assays and the Microtox® Acute Assay are frequently used for aqueous toxicity assessment, the project team compared toxicity data obtained with Shk1 and PM6 in batch-wise assays with the activated sludge and Microtox® data. The project team also included in the comparison data previously obtained with the continuous Shk1 assay (Ren, 2001).

3.3.1 Growth Media

Nutrient broth (NB) was used as the growth medium for Shk1 and PM6. This medium was prepared by dissolving nutrient broth (Fisher Scientific, Atlanta, Georgia) in de-ionized water to achieve a concentration of 8 g/L. NB medium was heat sterilized at 121°C and, after cooling, amended with 25 mg/L tetracycline before use. Because the bioreporters contained genes for resistance to tetracycline, this approach permitted the growth only of cells containing the plasmid.

3.3.2 Cell Preparation

Shk1 and PM6 were grown in 1 L flasks with 100 ml NB medium in an oscillating incubator at 30°C and 200 rpm. When optical density measured with a spectrophotometer at 600 nm (DU® 520, Beckman Coulter Inc., Fullerton, California) reached 1, cells were harvested and cell solutions were transferred to plastic centrifuge bottles. Cells were separated from growth medium by centrifugation (Beckman J2-21) at 14,000 × *g* and 4°C for 20 minutes (for 100 ml cell solution). The liquid supernatant was discarded after centrifugation. The cell pellets were resuspended in 4°C 0.1 mol/L KCl solution and were stored in a refrigerator at 4°C overnight.

3.3.3 Assay Toxicants

The toxicants involved in this research included eight metals and 48 organic compounds and were obtained from Sigma-Aldrich unless otherwise specified. A complete list of the metal and organic toxicants is provided in Table 3-1. For metals, $ZnCl_2$, $CuCl_2$, $PbCl_2$, $NiCl_2$, CrO_3, $Cr(H_2O)_6Cl_3$, $HgCl_2$, and a cadmium standard solution (cadmium in dilute nitric acid, Ricca Chemical Company, Arlington, Texas) were used. For organic chemicals, compounds with a wide variety of chemical structures were selected. These toxicants were chosen because they often appear in wastewaters. Stock solutions of the toxicants were prepared by dissolving toxicants in de-ionized water. Some phenols had low solubility, in which case a small amount of acetone was used to dissolve the compounds. In toxicity assays (described below), the research team prepared four successive two-fold dilutions of the stock solutions of assay toxicants achieve a total of five toxicant concentrations. If the 50% effective concentration (EC_{50}) was below the lowest tested concentration of a toxicant, the stock solution was further diluted with de-ionized water .

3.3.4 Assay Procedure

Refrigerated sensor cells were activated 20 minutes before toxicity assays in a water bath set at room temperature. Activated cells were mixed with each of the dilutions of toxicant solutions to achieve a total volume of 1 ml consisting of 0.1 ml cell solution and 0.9 ml toxicant solution. Also, a control was formed by 0.1 ml cell solution and 0.9 ml de-ionized water. Luminescence was measured seven minutes after mixing with either toxicant or de-ionized water. A Wallace Victor^2 Multilabel Counter (Perkin Elmer Life Sciences, Boston, Massachusetts) was used for luminescence measurements, which were performed at room temperature. Data were recorded and stored for further analysis only if the control luminescence did not shift by more than 5%.

3.3.5 Toxicity Data Acquisition

The toxicity endpoint involved was luminescence repression (*LR*), which was calculated according to Equation 3.3.5-1.

$$LR = \frac{L_c - L_t}{L_c} \qquad (3.3.5\text{-}1)$$

where L_c is luminescence of cells mixed with de-ionized water, and L_t is luminescence of cells mixed with toxicant. The research team conducted four replications for each of the toxicants . The results from the four replications were combined and *LR* was versus toxicant concentration was plotted to produce concentration-response curves. Toxicity data were reported as EC_{50} values (i.e., concentrations of toxicants that caused 50% luminescence repression). EC_{50} values were obtained by a least squares regression of the relatively linear portion of the concentration-response curves.

3.3.6 Toxicity Data Comparison

The investigators secured from the literature toxicity data in the form of EC_{50} values for the Microtox® and activated sludge respiration or growth inhibition assays . This information is shown in Table 3-1. In the case of metals, the concentrations presented in Table 3-1 are those that cause upsets in activated sludge process. However, these data were found in various studies in the literature, and what constituted an upset was defined differently by different researchers. These data also appeared in previously published work in which data sources were presented (Ren, 2001, Ren and Frymier, 2003a, 2003b). Toxicity data for the continuous Shk1 assay were obtained at UT and were reported previously (Ren and Frymier, 2002). Toxicity data obtained with Shk1 and PM6 in batch-wise assays were then compared with literature data.

To study the correlation structure displayed by the data, correlations between toxicity data obtained using Shk1 and PM6 in batch-wise assays, the continuous Shk1 assay, the Microtox® assay, and the activated sludge inhibition assays were inspected by examining the Pearson correlation coefficients for each pair of assays. The project team used JMP 5.0 (SAS Institute, Cary, North Carolina) for the statistical analyses.

3.4 Results

3.4.1 Toxicity Data Acquisition

Toxicity data evaluated with Shk1 and PM6 in the batch-wise assay protocol are presented in Table 3-1. Toxicity of the compounds under investigation spanned several orders of magnitude. For several compounds (e.g., methylene chloride in Table 3-1), 50% luminescence repression was not observed at the highest tested concentrations and therefore EC_{50} values were reported as greater than the highest concentration tested. For one or more of the following reasons, the investigators did not increase concentrations of test compounds to obtain exact EC_{50}: 1) the tested concentrations were already too high to be realized in WWTPs, 2) current data were sufficient to show sensitivities of the assays, and 3) some of the concentrations were at their solubility limits.

3.4.2 Toxicity Data Comparison

Toxicity data in Table 3-1 indicates that there is no general pattern regarding the relative sensitivities of the assays based on Shk1 and PM6. One assay may be slightly more sensitive than another assay for certain compounds but less sensitive for other compounds. Because of the large number of organic compounds and because the toxicity endpoints for the activated sludge inhibition assays were more uniform for organic compounds (either growth or respiration inhibition), the research team focused on the organic compounds below.

For organic compounds, toxicity data of the batch-wise Shk1 and PM6 assays were highly correlated. Figure 3-1 plots batch-wise Shk1 EC_{50} versus batch-wise PM6 EC_{50}.

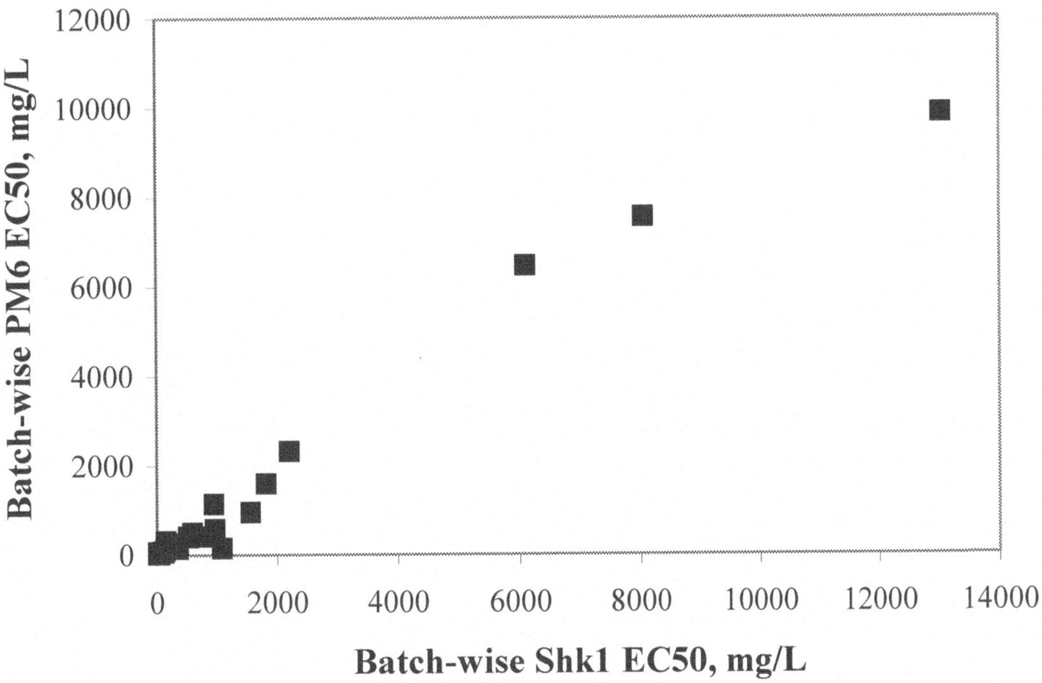

Figure 3-1. Correlation Between Toxicity Data of Batch-wise PM6 and Shk1 Assays.

A regression of batch-wise PM6 EC50 on batch-wise Shk1 EC50 produced the following equation with n (number of observation), r^2 (coefficient of determination), F (Fisher statistic) and the associated p value noted.

$$\text{PM6 EC}_{50}\ (\text{mg/L}) = 1.16\ \text{Shk1 EC}_{50}\ (\text{mg/L}) + 13.3 \qquad (3.4.2\text{-}1)$$

$$n = 34,\ r^2 = 0.97,\ F = 1130,\ p < 0.0001$$

A regression of batch-wise Shk1 EC_{50} on continuous Shk1 EC_{50} generated a statistically significant regression model ($p < 0.0001$), yet the quality of fit was significantly poorer than that of Equation 3-2, with an r^2 of only 0.78. Because data of the batch-wise Shk1 and PM6 assays were correlated, regression of batch-wise PM6 EC_{50} on continuous Shk1 EC_{50} showed a similar pattern with an r^2 of 0.86, still lower than that of Equation 3-2.

The investigators further studied correlation structure of toxicity data by considering the Microtox® and activated sludge inhibition assays in addition to the assays based on Shk1 and PM6. As seen in Table 3-1, data availability was an issue; some Microtox® and activated sludge inhibition data were not available for certain compounds. To ensure a fair comparison, such compounds were excluded in the analysis. As a result, only 18 organic compounds with data for all five assays were used in the analysis. Pair-wise correlation coefficients between assays are displayed in Table 3-2. These results indicate that, for the compounds under investigation, correlation existed between toxicity data of every pair of the assays. If the activated sludge inhibition assay was used as a reference assay, batch-wise PM6 assay exhibited the strongest correlation with it.

Figure 3-2 shows the $logEC_{50}$ data of the batch-wise PM6 and Shk1 and the Microtox® assays versus the $logEC_{50}$ data of the activated sludge inhibition assays. Logarithm transformation of the original data was performed to facilitate presentation of the figure. The same conclusion inferred from Table 3-2 is evident in Figure 3-2: Batch-wise PM6 assay correlated the best with activated sludge inhibition assay. In addition to correlation, Figure 3-2 suggests that, because the points representing both the batch-wise PM6 and Shk1 assays were generally close to the 45° diagonal line, these two assays had sensitivities similar to that of activated sludge inhibition assays. By contrast, the Microtox® assay is generally more sensitive than the activated sludge inhibition assay, which has been previously reported in the relevant literature, because many of the points representing the Microtox® assay were considerably below the 45° line. A figure replacing batch-wise PM6 and Shk1 assays with the continuous Shk1 assays had previously been shown (Ren and Frymier, under revision).

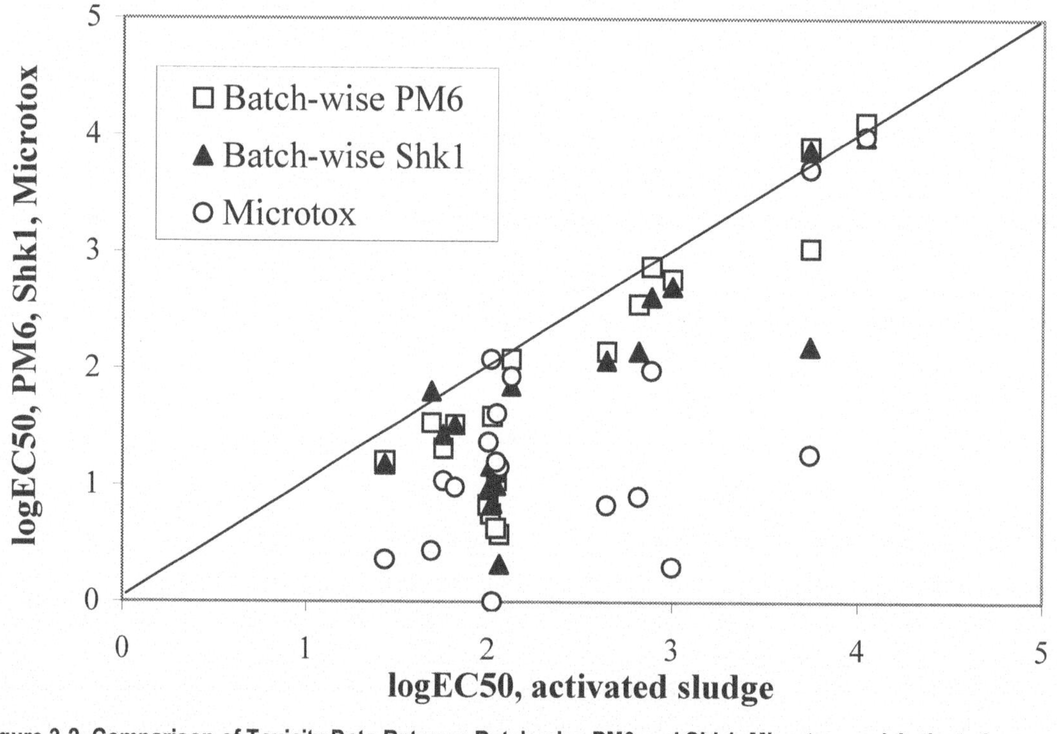

Figure 3-2. Comparison of Toxicity Data Between Batch-wise PM6 and Shk1, Microtox, and Activated Sludge Inhibition Assays.

3.5 Discussion

The purpose of replacing the growth media with the KCl solution before storage at 4°C was three-fold. The first purpose was to create a uniform assay environment for all organisms used in the battery. Second, while the growth medium for *Hyphomicrobium* species (data shown in Appendix B) was transparent, the NB medium used for *Pseudomonas* species had a dark color. The interference of color to luminescence-based toxicity assays has been pointed out (Lappalainen et al., 2001). The project team believed it was advantageous to completely eliminate the effect of color by replacing the growth medium with KCl solution. The third reason for replacement was that it is difficult to determine the potential interference caused by such undefined constituents in complex media (such as NB) on toxicity assay.

Studies in the literature have shown that each species and test procedure has its own sensitivity pattern to toxicants (Dutka and Kwan, 1981, 1982, Dutka et al., 1983). It is generally believed that a battery of assays instead of a single assay would give a better assessment of toxicity. A test battery might include organisms of different trophic levels (Blaise, 1991, Castillo et al., 2000, Ribo, 1997). For example, Blaise (1985) suggested using four test organisms of increasing levels of biological organization (luminescent bacteria, crustacea, alga, and fish) to assess the toxic effect of wastewater on the environment. The use of test batteries for toxicity assessment has been documented in the literature (e.g., Castillo et al., 2000, Fochtman et al., 2000, Kahru et al., 2000). However, appropriate and efficient selection of assays as test battery components has not been well investigated. Ren and Frymier (2003c) made one of the first attempts to identify methods for choosing assays in a test battery for wastewater toxicity assessment. Those authors also reviewed general considerations to bear in mind when choosing assays as test battery components. Two key considerations in selecting test battery components are that the components be representative of the environment or system of interest and that they not provide redundant toxicity information.

PM6 and Shk1 could be good candidates as sensor organisms due to their luminescence levels. Given the correlation with results of the activated sludge inhibition assays, either of the batch-wise PM6 and Shk1 assays could be a suitable assay for wastewater toxicity effects on activated sludge respiration (Table 3-2 and Figure 3-2). Due to their origin, both PM6 and Shk1 can be considered representative of at least part of the activated sludge environment. Using the batch-wise PM6 and Shk1 assays in a test battery clearly satisfies the requirement of representativity. However, including these two assays based on these organisms may not be appropriate because their results are highly correlated, at least for the single compounds tested, meaning that toxicity information obtained from one of these two assays would be redundant. This redundancy in toxicity information could arise because both PM6 and Shk1 belong to *Pseudomonas* species. However, it is worth noting that strong correlations do not exist only between organisms of the same species; they can even exist between toxicity data of very different species, e.g., correlation between toxicity data for protozoa *Tetrahymena pyriformis* and fathead minnow *Pimephales promelas* (Schultz et al., 1998) and between higher plants *Cucumis sativus* and *T. pyriformis* (Wang et al., 2002).

In light of the strong correlation between toxicity data of the batch-wise PM6 and Shk1 assays, a closer examination of the sensitivity pattern of the two assays is needed. This is because if one assay is particularly sensitive to a certain class of compounds (e.g., amines, phenols, etc.), it may still be included in a test battery despite the overall correlation of toxicity data. However, this is not the case for the batch-wise PM6 and Shk1 assays, as results in Table 3-1 indicate that neither is particularly sensitive to any class of compounds. This is consistent with the slope in Equation 3-2 being near unity.

Luminescent organisms that belong to other species, such as *Hyphomicrobium* and *Nitrosomonas*, which could also be representative of parts of the activated sludge community, might generate toxicity data not correlated with those of the *Pseudomoans* species. The project team created three luminescent *Hyphomicrobium* strains. Unfortunately, these strains do not produce luminescence at an acceptable level in the toxicity assay used to warrant extensive investigation at this time.

In this research, the project team only determined the toxicity of single compounds in this research. The team acknowledges that actual wastewater represents a more complicated situation. Wastewater composition and constituent concentrations are frequently highly variable. Nonetheless, wastewater toxicity, if any, results almost inevitably from a mixture of chemical compounds. It is widely known that there are interactive effects (i.e., synergistic, antagonistic, and additive) between toxicants that exist in mixtures. Consequently, the toxicity of a mixture cannot necessarily be inferred from the individual toxicity of the mixture components (Mowat and Bundy, 2002).

The interactive effects between mixture components may have at least two implications. First and foremost, toxicity of certain mixtures needs to be evaluated if such mixtures can reasonably be expected to appear frequently in wastewater. While some well-established models exist for mixture toxicity (especially for mixtures of organic compounds) such as concentration addition (Frei, 1913) and independent action (Bliss, 1939), in many situations toxicity of mixtures has to be experimentally determined. For example, Cooper (2001) examined the toxicity of metals in the presence of organic solvents due to the wide use of solvents in industries. The project team did not conduct toxicity assays for mixtures because of the lack of reported data in the literature for the Microtox and activated sludge inhibition assays that would be a basis for comparison purposes. The second implication of mixture toxicity is that the correlation between toxicity data for single compounds of the batch-wise PM6 and Shk1 assays may or may not be readily generalized to mixtures of toxicants. However, due to the similarities between PM6 and Shk1 as mentioned before, it is unlikely that the interactive effects of toxicant mixtures are different for these two organisms. Thus, it should not be surprising if the batch-wise PM6 and Shk1 assays also have correlated results for toxicity of mixtures.

The above discussion also applies to the continuous Shk1 assay because Table 3-2 shows that results of the continuous Shk1 assay are also correlated with results of the batch-wise PM6 and Shk1 assays. In any event, batch-wise PM6 and Shk1 and continuous Shk1 assays provide redundant results of toxicity assessment, and the Microtox® assay provides results that are more poorly correlated with activated sludge respirometry than these three assays. Forming a test battery for wastewater toxicity assessment that contains more than one of the new assays is therefore not advisable since there is likely to be a high degree of data redundancy.

Table 3-1. Toxicity Data (EC_{50}, mg/L). Standard Errors, Where Available, are Shown in Parentheses.

	Batch-wise PM6[1]	Batch-wise Shk1[1]	Continuous Shk1[2]	Microtox[5]	Activated sludge inhibition[7]
Metals					
Cr^{3+}	140 (11)	311 (14)	303[3]	10.7	-
Cr^{6+}	22 (0.2)	2 (0.05)	38.9[3]	-	0-11[8], 25[9]
Cd^{2+}	163 (4)	384 (8)	1.9[3]	98	12.5[10]
Cu^{2+}	3.1 (0.1)	0.6 (0.05)	9.4[3]	0.72-2.46	10[11], 29[12]
Hg^{2+}	1.4 (0.3)	0.1 (0.002)	0.9[3]	0.064	4.5[12]
Pb^{2+}	27 (2)	0.4 (0.06)	3.8[3]	2.56	350[12]
Ni^{2+}	0.2 (0.1)	0.9 (0.08)	96[3]	917-1056	4.3[12], 10[11]
Zn^{2+}	0.6 (0.05)	0.2 (0.01)	1.9[3]	13.8-55.5	1.2[12], 10[11]
Organics					
Benzene	588 (14)	509 (14)	202	2.0	993
Nitrobenzene	342 (19)	262 (25)	203	33	-
Ethylbenzene	>117	94 (4)	57	5.5	222
Ether	> 642	> 642	3536	-	-
Ethyl acetate	8051 (542)	7556 (531)	1711	5160	5427
Cyclohexanone	1081 (26)	157	117[4]	18.5	5452
1-Butanol	6100 (306)	6462 (149)	2745	2056	-
1-Pentanol	> 2201	938 (127)	737	298	3528
1-Propanol	13033 (514)	9861 (1018)	8303	9862	10875
Chloroform	714 (60)	> 2023	701	441	-
1,1,1-Trichloroethane	357 (14)	142 (11)	49	8.04 (15 min)	659
Glutaldehyde	80 (7)	79 (6)	167	75.7	-
Methylene chloride	> 11930	> 11930	2935	1878	1994
4-Chlorobenzaldehyde	53 (3)	27 (2)	47	10.2	-
1,2-Dichlorobenzene	34 (5)	64 (2)	81	2.74	49
Formaldehyde	148 (10)	316 (13)	36	3.0	-
2,4-Dinitrotoluene	> 135	129 (8)	242	33	-
Bromoform	529 (31)	410 (55)	165	-	-
Benzyl alcohol	1557 (75)	966 (54)	635	63.7	-
p-Xylene	> 116	42 (2)	57	5.7 (30 min)	166

1,2,3-Trichloropropane	518 (26)	418 (14)	211	18.7	-
1,5-Dichloropentane	> 110	105 (6)	59	15.7	-
1,4-Dichlorobutane	965 (42)	582 (162)	270	61.5	-
n-Butylamine	11 (1)	13 (0.7)	318	41.1	111
Trichloroethylene	757 (31)	414 (27)	190	96.8	770
Cyclohexylamine	6 (0.4)	15 (0.6)	407	120 (30 min)	103
Bromochloromethane			3228	-	2672
Diethylamine	7 (0.3)	9 (0.8)	327	23.4	100
Cyclohexane	120 (6)	71 (16)	95	85.5	133
Octane	> 127	> 127	163	730	3025
n-Amyl acetate	8 (0.5)	74 (6)	183	-	-
Isobutyl acetate	1819 (112)	1601 (84)	294	-	2156
1,2-Dibromoethane	2204 (55)	2323 (223)	1409	735	-
Methyl ethyl ketone	> 3623	> 3623	629[4]	14	1873
Ethanolamine	4 (0.2)	2 (0.3)	560	14.3	115
m-Cresol	141 (5)	118 (13)	189	6.8	440[13]
2,3,4,5-Tetrachlorophenol	0.2 (0.1)	1 (0.2)	7.3	0.3	-
3-Phenylphenol	18 (1)	13 (1)	24	0.1[6]	-
2,4-Dinitrophenol	4 (0.5)	10 (0.6)	105	15.8	110[14]
4-Nitrophenol	21 (0.9)	27 (0.8)	43	10.9	57[15]
1-Hexanol	947 (86)	1149 (38)	202	21.5	-
2,6-Dichlorophenol	33 (0.8)	33 (0.7)	120	9.6	66[15]
4-Chloro-3-methylphenol	4 (0.3)	5 (1)	18	0.3	-
4-Chlorophenol	39 (5)	7 (5)	20	1.0	105[15]
2,4-Dichlorophenol	16 (0.9)	15 (0.7)	30	2.3	27[15]
1-Naphthol	7 (0.7)	27 (2)	53	3.7	-
2,4,5-Trichlorophenol	1 (0.2)	3 (0.1)	13	1.2	23[13]

[1]: Numbers in parentheses are standard errors of the corresponding EC50 values.
[2]: Ren and Frymier (2002) unless otherwise specified.
[3]: Ren and Frymier (2003a).
[4]: Ren and Frymier (2003b).
[5]: Kaiser and Devillers (1994) and 5-min data unless other specified.
[6]: Calculated from data in Schultz et al. (1998).
[7]: Sun et al. (1994) unless otherwise specified.
[8]: Mazierski (1995).
[9]: Gokcay and Yetis (1991).
[10]: Estimated from data in Madoni et al. (1999).
[11]: Yetis and Gokcay (1989).
[12]: Dutka and Kwan (1984).
[13]: Blum and Speece (1991).
[14]: Volskay and Leslie Grady (1988).
[15]: Walker (1989).

Table 3-2. Correlations Between Batch-wise PM6, Batch-wise Shk1, Continuous Shk1, Microtox, and Activated Sludge Inhibition Assays for Selected Compounds in Table 3-1.

	Batch-wise PM6	Batch-wise Shk1	Continuous Shk1	Microtox	Activated sludge inhibition
Batch-wise PM6	1				
Batch-wise Shk1	0.99	1			
Continuous Shk1	0.92	0.88	1		
Microtox	0.99	0.98	0.95	1	
Activated sludge inhibition	0.93	0.89	0.87	0.90	1

CHAPTER 4.0

CONCLUSIONS AND RECOMMENDATIONS

4.1 Creation of New Biosensors

Six new bioluminescent toxicity bioreporters were constructed. Three are *Pseudomonas* spp. isolated from a municipal activated sludge wastewater treatment plant and three are *Hyphomicrobium* sp. originally isolated from an industrial wastewater treatment plant. The *Pseudomonas* spp. bioreporters (PM2, PM5, and PM6) grew much faster in batch culture than the *Hyphomicrobium* sp. bioreporters (H1, H2, H3). For all species, light was proportional to cell density.

The work presented in this report is of long-term interest to the project team and continues. However, attempts to develop additional bioreporters using some of the activated sludge species to provide unique information have not yet succeeded. The use of conjugation to transfer the *lux* genes is hindered by the lack of established selective media or culture conditions for some of these species. In many of the activated sludge strains tested, electroporation over a range of conditions was not successful in introducing the *lux* genes.

4.2 Screening of the New Biosensors

The new biosensors created in the first phase of this project were then tested to determine which biosensors would be good candidates to include in an array of biosensors for influent toxicity screening. In preliminary screening, the new strain PM6 was shown to produce sufficient light to warrant further screening. The remaining strains, H1, H2, H3, PM2, and PM5, did not produce sufficient light at what the project team considered reasonable cell densities to warrant further screening with the assay protocol they used.

The project team evaluated the toxicity of eight metals and 48 organic compounds with the batch-wise toxicity assay protocol based on luminous bacteria PM6 and Shk1 (Shk1 was a previously constructed biosensor), both of which are *Pseudomonas* strains. Results of the toxicity assessment indicated that data obtained from these two assays were highly correlated and had approximately the same sensitivity to all classes of chemical compounds in the collection of toxicants tested.

Comparing the results of the batch-wise PM6 and Shk1 assays and those of the continuous Shk1 (developed and published previously), Microtox®, and activated sludge inhibition assays, the project team found that toxicity data of all these assays were correlated; however, the batch-wise PM6 and Shk1 assays had responses to toxicants most similar to the activated sludge inhibition assays. Based on the results of this research, the project team concluded that, while an array of sensors either the Shk1 or PM6 assays would be useful, using both of the organisms would not yield additional information. Therefore, continued work is needed to develop appropriate sensor organisms/assays. In the construction of new biosensors, the first phase of the work indicated that further efforts are needed to develop genetic tools tailored to the types of organisms found in wastewater treatment systems.

4.3 Recommendations

Based on the results of this project, the research team has identified several potential areas for future research in the short and long term and some suggestions to future researchers in this area. These are summarized below and are categorized as short- and long-term research needs.

Short-term research should focus on two tasks. First, research should focus on obtaining a nitrifying biosensor, either by more focused effort to insert the available plasmid (Iizumi, T., M. Mizumoto, and K. Nakamura. Appl. Environ. Microbiol. 64(10): 3656-3662, 1998) in a *Nitrosomonas* bacterium or by obtaining the biosensor from the only group known to have maintained a bioluminescent nitrifying bacterium (Ludwig C, S. Ecker, K. Schwindel, H.G. Rast, K.O. Stetter, and G. Eberz, Archives of Microbiology, 172 (1): 45-50, 1999). Second (and in parallel with the first item), research should focus on developing an assay that has the sensitivity and ability to accommodate the low growth rate required to utilize the *Hyphomicrobium* biosensor developed as part of this project. The value of the *Hyphomicrobium* biosensor was not completely understood until the research team realized the difficulty of generating new biosensors that were not *Pseudomonads*. Until the latter phase of the project, it was not known that *Pseudomonads* would exhibit such highly correlated toxicity responses d. With these two organisms (a *Nitrosomonas* and a *Hyphomicrobium*) in addition to PM6 or Shk1, it would be possible to test the central hypothesis of the original project; that is, additional biosensors can be created that give unique information about the toxicity of a sample.

Long-term research should focus on developing more effective methods for getting the *lux* genes into wastewater treatment plant bacteria and culturing populations of slow-growing wastewater treatment plant bacteria. Many wastewater treatment plant bacteria have protective sheaths on their filaments or exopolysaccharide layers on their exteriors. The research team observed these make it harder to get genes across the membrane of the target cells. In earlier work (data not shown), chemical mutagenesis was necessary to create a mutant strain without an exopolysaccharide layer before the target cells would take up a transposon vector. Chemical mutagenesis is a drastic method and can result in unintended mutations; other methods of breaking up the exopolysaccharide layers should be pursued first (including possibly sonication or mild alkali treatment before electroporation). In genetic transformation research, it is also always a possibility that the bioluminescence genes and associated enzymes simply will not function or that the plasmids cannot be replicated in the host cells. Many wastewater treatment plant bacteria are slow-growing microorganisms that do not grow well or cannot be cultured using current standard laboratory techniques. Simply using spread plate and conjugation methods with a consortia of wastewater treatment plant bacteria and *E. coli* carrying the *lux* genes on a self-replicating plasmid will therefore result most frequently in bioreporters that are fast growing microorganisms present in high numbers like *Pseudomonads*. This is why pure cultures were used in the transformation efforts documented in this report. The pure culture transformation efforts of the research team further indicate that, unlike *Pseudomonads*, some wastewater treatment plant microorganisms cannot maintain the self-replicating plasmid pUTK2 and must be transformed with transposon vectors that allow the genes to be inserted in the chromosomal DNA.

APPENDIX A

METHODS FOR CREATION OF THE BIOREPORTERS

A.1 Media

Several types of media were used in this study, and each medium is described in the sections below.

A.1.1 Isolation Medium for *Hyphomicrobium* (IMH)

IMH medium is only somewhat selectable for *Hyphomicrobium* strains. It was prepared by adding the following ingredients into 100 mL of distilled water: KH_2PO_4, 0.136 g; $(NH_4)_2SO_4$, 0.05 g; $CaCl_2 \cdot 2H_2O$, 0.001 g; $MnSO_4 \cdot 4H_2O$, 0.00031 g; $Na_2HPO_4 \cdot 4H_2O$, 0.213 g; $MgSO_4 \cdot 7H_2O$, 0.02 g; $FeSO_4 \cdot 7H_2O$, 0.0005 g; and $NaMoO_4 \cdot 2H_2O$, 0.00025 g. The pH of the medium was adjusted to 7.2-7.4. After autoclaving at 121°C for 15 minutes the medium was supplemented with 1 ml of filter-sterilized methylamine•HCl stock solution containing 1 g methylamine•HCl per mL of distilled water. Solid versions of the medium contained 2% agar.

A.1.2 Czapek-Dox Medium (CD)

Gordonia amarae was grown on Czapek-Dox broth (Difco, Becton Dickinson, Sparks, MD). This medium contains the following ingredients per liter: saccharose 30. 0 g; sodium nitrate 3.0 g; dipotassium phosphate 1.0 g; magnesium sulfate 0.5 g; potassium chloride 0.5 g; and ferrous sulfate 0.01 g. The pH of this medium is 7.1-7.5. Solid versions of the medium contained 2% agar.

A.1.3 Rouf and Stokes Medium (Zoo)

The growth medium for both *S. natans* and *Z. ramigera* was modified Rouf and Stokes medium. Modified Rouf and Stokes medium (Rouf et al., 1964) contained the following ingredients per liter: peptone, 5.0 g; $MgSO_4 \cdot 7H_2O$, 0.2 g; $Fe(NH_4)(SO_4)_2$, 0.15 g; sodium citrate, 0.1g; $CaCl_2$, 0.05g; $MnSO_4 \cdot H_2O$, 0.05; $FeCl_3 \cdot 6H_2O$, 0.01 g. The pH of the medium was adjusted to 7.2-7.4 with NaOH. After autoclaving at 121°C for 20 minutes, the medium was supplemented with 1 mL of filter-sterilized vitamin stock solution containing 5 mg vitamin B_{12}, 4 mg biotin, and 4 mg of thiamin. Solid versions of the medium contained 2% agar.

A.1.4 P Medium

For culturing *Nitrosomonus europaea*, P medium was made by adding 100 mL of autoclaved solution B (pH=8.0), which contained 27.2 g KH_2PO_4 and 2.4 g Na_2HPO_4 per 500 mL of distilled water into 900 ml of autoclaved solution A, which contained the following ingredients per liter of distilled water: $(NH_4)_2SO_4$, 3.3 g; KH_2PO_4, 0.41 g; 1 M

Mg SO_4 stock solution, 0.75 mL; 1 M $CaCl_2$ stock solution, 0.2 mL; 30 mM $FeSO_4$/50 mM EDTA stock solution, 0.33 mL; 50 mM $CuSO_4$ stock solution, 0.01 mL. Then 8 mL of filter sterilized 5% Na_2CO_3 (anhydrous) was added into the medium.

A.1.5 PLB Medium

PLB medium was the solid medium for culturing *N. europaea*. The medium contains two parts (based on 1 liter). The first part includes 500 mL water; tryptone, 5 g; yeast extract, 2.5 g; KH_2PO_4, 1.1 g; Na_2HPO_4, 6 g; gellan gum, 15 g; pH 7.6. The second part includes 450 mL water. The two parts were autoclaved separately. After sterilization, stock solutions including 100 g/L $(NH_4)_2SO_4$, 12.5 mL; 5 g/L $CaCl_2$, 0.5 mL; 10 g/L $MgSO_4$, 0.5 ml; and 1g/L Fe-EDTA, 0.5 mL were added immediately into the second part. Part two was then added to part one. The solidifying temperature of gellan gum is higher than that of agarose, so pouring the plates has to be executed quickly to avoid premature congealing to the solid state.

A.1.6 Glucose and Yeast Extract Medium (ATCC Medium: 1858)

The growth medium for *Z. ramigera* contained the following ingredients per liter: yeast extract, 2.5 g; peptone, 2.5 g. The pH of the medium was adjusted to 7.0-7.4 with NaOH. After autoclaving at 121°C for 20 minutes, the medium was supplemented with 1 mL of filter-sterilized vitamin stock solution (the same as used in Zoo medium). Solid versions of the medium contained 2% agar.

A.1.7 Luria-Bertani Broth (LB Broth)

LB broth (Difco, Becton Dickinson, Sparks, MD) is suitable for cultivation of microorganisms such as *Escherichia coli* and *Pseudomonas spp.* It contains the following ingredients per liter: pancreatic digest of casein, 10.0 g; yeast extract, 5.0 g; sodium chloride 10.0 g. The pH of the medium is 6.8 - 7.2. Solid versions of the medium contained 2% agar.

A.1.8 *Pseudomonas* Isolation Agar

Pseudomonas Isolation Agar (Difco, Becton Dickinson, Sparks, MD) is commonly used for the isolation and identification of *Pseudomonas* strains. The ingredients of the medium are as follows, which are based on one liter: peptone, 20 g; magnesium chloride, 1.4 g; potassium sulfate, 10.0 g; irgasan, 0.025 g; glycerol, 20 ml; agar, 13.6 g. The pH of the medium is 6.8 - 7.2.

A.1.9 Nutrient Broth

Nutrient Broth (Sigma, St. Louis, MO) contains beef extract and peptone and is a general growth medium suitable for cultivation of microorganisms such as *Escherichia coli* and *Pseudomonas spp.* 8 g of premixed powder were dissolved into one liter of distilled water and autoclaved. Solid versions of the medium contained 2% agar.

A.2 Construction Methodologies

A.2.1 Methodology for Construction of *Hyphomicrobium* str. M3 Bioluminescent Toxicity Bioreporters

E. coli DH5α Tn5 *luxAB*, *E. coli* DH5α Tn4431, and *E. col*i S17-λpir with pUTK2 were streaked on LB agar medium with 10 mg/L tetracycline and cultured at 37°C overnight. *Hyphomicrobium* sp. M3 was grown on *Hyphomicrobium* isolation medium (IMH) at room temperature. Individual colonies of *H.* str. M3 were picked to IMH broth, and *E. coli* cultures were picked to LB medium with 10 mg/L tetracycline. After growth of all cultures, 100μL of each culture were spread on nutrient broth (NB) agar plates. Fifty μL of each *E. coli* strain together with 50 μL *Hyphomicrobium* were also spread plated onto NB agar plates. The plates were incubated at room temperature for two days. The cells were collected by scraping with an inoculation loop and resuspended in 250 μL PBS. Two plates were inoculated with different volumes of cell solution, 100 μL and 10 μL, by spreading the solution onto IMH agar medium with 25 mg/L tetracycline and incubated at room temperature for two weeks. Individual colonies on the plates were picked and streaked as lines (50 lines/plate) on IMH agar medium with tetracycline and incubated at room temperature. Strains producing visible light were detected in a dark room and restreaked to IMH agar medium with 25 mg/L tetracycline. Morphologies of the cells in resulting colonies were checked microscopically.

The three highest light-producing strains identified were designated as *Hyphomicrobium* Tn4431 H1, H2, and H3. These three stains were grown in 100 mL IMH medium with 10 mg/L tetracycline. Another 100 mL IMH medium with 10 mg/L tetracycline was prepared as a control and was not inoculated. Every six hours the optical density in each culture was measured by spectrophotometer, and the bioluminescence was measured by luminometer.

A.2.1 Methodology for Construction of *Pseudomonas* spp. Bioluminescent Toxicity Bioreporters

E. coli DH5α Tn5 *luxAB*, *E. coli* DH5α Tn4431 and *E. col*i S17-λpir pUTK2 were grown on LB agar medium with 10 mg/L tetracycline at 37°C overnight. Four different tetracycline-sensitive *Pseudomonas* strains (PM2, PM4, PM5, and PM6) previously isolated from activated sludge were grown on *Pseudomonas* isolation agar (PIA) at 30°C overnight. The cells were collected by scraping and were resuspended separately in 250 μL phosphate buffered saline (PBS). Each of the *Pseudomonas* strains (50 μL) was mated with each of the *E. coli* strains (50 μL) on NB agar plates. Control matings consisted of each *E. coli* and *Pseudomonas* strain alone. The plates were incubated at room temperature for two days. After mating, the cells were collected by scraping with an inoculation loop and resuspensed in 250 μL PBS. One hundred μL and 10 μL of each mating cell solution were plated onto PIA agar medium with 25 mg/L tetracycline, and were incubated at 30°C for two days. Large, well-separated colonies on the plates were picked and streaked as lines (50 lines/plate) on PIA agar medium with tetracycline, and incubated at 30°C. Strains producing visible light were detected in a dark room and restreaked to PIA agar medium with 25 mg/L tetracycline.

Strains PM2 pUTK2 and PM5 pUTK2 were chosen for further study. These two stains were grown in 100 mL NB medium with 10 mg/L tetracycline. Another 100 mL NB medium with 10 mg/L tetracycline was prepared as an uninoculated control. The optical densities and bioluminescence of samples from each culture were measured at approximately six-hour intervals.

A.3 Methodology for Electroporation into *Sphaerotilus natans* and *Zoogloea ramigera*

A.3.1 Preparing Electrocompetent Cells

The recipient strains were grown until early logarithmic phase. *S. natans* was grown in Zoo medium at 30°C for three days. *Z. ramigera* was grown in 1858 medium at 30°C for three days. The cells were cooled with ice and harvested by centrifugation at 4500 rpm for 20 minutes. The cell pellets were washed three times with a 10% v/v glycerol water solution. The cells were kept cold during the entire procedure. The cells were resuspended in approximately the same volume of cold electroporation buffer (10% glycerol water solution).

A.3.2 Plasmid DNA Preparation

The plasmid DNA pSa325: Tn4431 for electroporation was extracted from strain *Escherichia coli* DH5 α Tn4431. The culture was grown overnight in LB medium with 10 mg/L tetracycline, and then the plasmid DNA was extracted using the S.N.A.P Invitrogen MidiPrep Kit (Carlsbad, CA). Ten μL of the DNA solution were run on a 0.8 % electrophoresis gel to check the size, and the concentration of DNA was determined using a spectrophotometer (OD_{260}). The rest of the DNA solution was partially digested with *Pst*I and stored at -80°C.

A.3.3 Electroporation of Cells

The electroporator voltage was set at 2.5 kV, minimum resistance at 200 Ω and maximum resistance at ∞ Ω. The field strength was 1.25 kv/mm gap. Ten μL of linearized plasmid DNA (1.6 μg of DNA) were added into tubes containing 100 μL electrocompetent cells. The mixture of cells and DNA was transferred to a pre-chilled 2 mm gap cuvette. The outside of the cuvette was dried. The cuvette was inserted into the cuvette holder and placed into the electroporator. The electrical pulse was applied, and then the cells and DNA were mixed with 1 mL of growth medium (Zoo medium for *S. natans*; 1858 medium for *Z. ramigera*). Both strains were incubated at 30°C for 12 hours. After incubation, all the cells were plated on solid medium (Zoo medium for *S. natans*; 1858 medium for *Z. ramigera*) with 25 mg/L tetracycline for selection. Another set of cells was prepared in parallel as a no-insert control. When colonies appeared on the experiment plate (with DNA added), they were restreaked on another agar plate with 25 mg/L tetracycline and incubated at 30°C. After seven days, individual colonies from each strain were inoculated into 1 mL of liquid medium with 1μL of *n*-decyl aldehyde, and the bioluminescence was measured at 60, 120, and 180 minutes. The five highest light-producing isolates were regrown in liquid medium, and cell density and bioluminescence were measured with time.

APPENDIX B

METHODS FOR GENERATING TOXICITY PROFILES

B.1 Introduction

Members of the project team at Syracuse University created several luminescent sensor organisms and transferred the most promising ones to UT. There were three *Pseudomonas* strains denoted PM2, PM5, and PM6, and three *Hyphomicrobium* strains denoted H1, H2, and H3. Chapter 2.0 of this report describes the project team's creation of these organisms. Chapter 3.0 describes the team's toxicity testing efforts to determine the best strains to use in an array of biosensors. The testing in Chapter 3.0 involved only the strains PM6 and the previously created Shk1 because preliminary experiments documented in this Appendix indicated that the strains H1, H2, H3, PM3, and PM5 did not produce sufficient light in the final testing protocol to give reliable toxicity data. These preliminary experiments used a cell-to-toxicant mixing ratio of 9:1 (v/v) in the assay procedure. More often, assays are conducted with cell solution and test sample solution at a ratio of 1:1 (e.g., Cooper, 2001, Scheers et al., 2002) or 1:9 (e.g., Lajoie et al., 2002). This Appendix documents these preliminary experiments using the high (9:1) cell-to-toxicant mixing ratio. After these preliminary experiments were conducted and the strains H1, H2, H3, PM3 and PM5 eliminated from further consideration, it was possible to adopt a more dilute mixing ratio of 1:9 (v/v) in the batch-wise pure component assay protocol, which was described and discussed in Chapter 3.0.

B.2 Methods

The methodology the project team used for generating the toxicity data described here was essentially the same as that described in Chapter 3.0. The only difference was the mixing ratio between cell solution and toxicant solution. However, for completeness and convenience, the methodology is recounted in detail again below. As mentioned above, six recently created sensor organisms, i.e., PM2, PM5, PM6, H1, H2, and H3 were involved in this part of the research.

B.2.1 Cell Preparation

Nutrient broth medium (NB) was used as the growth medium for *Pseudomonas* strains, and for *Hyphomicrobium* strains, the *Hyphomocrobium* isolation medium (IMH) was used (see Appendix A for compositions). PM2, PM5, PM6, H1, H2, and H3 were grown in 1 L flasks with 100 ml appropriate medium in an oscillating incubator at 30°C and 200 rpm. When optical density measured with a spectrophotometer (DU® 520, Beckman Coulter Inc., Fullerton, California) at 600 nm reached 1.0, cells were harvested and cell solutions were transferred to plastic centrifuge bottles. Cells were separated from growth medium by centrifugation at 14,000 × *g* and 4°C for 20 minutes (for 100 ml cell solution). The liquid supernatant after centrifugation was discarded. The cell pellets were resuspended in 4°C 0.1 mol/L KCl solution to their original OD of 1.0. The resuspended cells were stored at 4°C overnight.

B.2.2 Assay Toxicants

The toxicants involved in this research included both metals and organic compounds and were obtained from Sigma-Aldrich unless otherwise specified. A complete list of the metal and organic toxicants is provided in Table B-1. For metals, $ZnCl_2$, $CuCl_2$, $PbCl_2$, $NiCl_2$, CrO_3, $Cr(H_2O)_6Cl_3$, $HgCl_2$, and a cadmium standard solution (cadmium in dilute nitric acid, Ricca Chemical Company, Arlington, Texas) were used. For organic chemicals, compounds with a wide variety of chemical structures were employed. These toxicants were chosen because of their frequent appearance in wastewaters. All chemicals were sufficiently pure, and further purification was not needed. Stock solutions of the toxicants were prepared by dissolving toxicants in de-ionized water. Some phenols had low solubility, in which case acetone was used as the carrier. In toxicity assays (described below), successive two-fold dilutions of the stock solutions of assay toxicants were prepared to achieve suitable toxicant concentrations. Four diluted solutions were used, in addition to the original solution of each toxicant were used. If the 50% effective concentration (EC_{50}) was below the lowest tested concentration of a toxicant, the stock solution was further diluted with deionized water.

B.2.3 Assay Procedure

Refrigerated sensor cells were activated in a waterbath set at room temperature for 20 minutes before toxicity assays. Activated cells were mixed with toxicant solutions to achieve a total volume of 1 ml, consisting of 0.9 ml cell solution and 0.1 ml toxicant solution. A control consisting of 0.9 ml cell solution and 0.1 ml de-ionized water was also made. Luminescence was measured seven minutes after mixing with either toxicant or de-ionized water. A Wallace Victor^2 Multilabel Counter (Perkin Elmer Life Sciences, Boston, Massachusetts) was used for luminescence measurements, which were performed at room temperature. Data were recorded and stored for further analysis only if the control luminescence did not shift by more than 5%.

B.2.4 Toxicity Data Acquisition

Luminescence repression (*LR*) was calculated according to Equation B.2.4-1.

$$LR = \frac{L_c - L_t}{L_c} \qquad \text{(B.2.4-1)}$$

where L_c is luminescence of cells mixed with de-ionized water, and L_t is luminescence of cells mixed with toxicant. Four replications for each of the toxicants were conducted. The results from the four replications were combined, and *LR* was plotted versus toxicant concentration to produce concentration-response curves. EC_{50} values were obtained by a least squares regression of the relatively linear portion of the concentration-response curves.

B.3 Results

Results indicated that HM3 did not produce light at an acceptable level to justify efforts to create toxicity data for this organism. Toxicity data obtained for the other five sensor organisms are presented in Table B-1 (for H1 and H2, only metals were tested, as it was quickly realized that these strains produced a light signal too near the instrument background to be reliable). As noted in Table B-1, 50% *LR* was not observed at the highest concentrations tested for many toxicants, resulting in a number of "censored" data. For some of the toxicants, the test concentrations were below their solubility limits, and their concentrations could be increased in the stock solution. However, this was not done because the purpose of this research was to develop assays for wastewater toxicity assessment. The censored data in Table B-1 represent toxicant concentrations already so high that they are unlikely to be observed in wastewaters.

B.4 Discussion

The mixing ratio between cell solution (nine parts) and toxicant solution (one part) described in B.2 Methods is not frequently seen in the relevant literature. The 9:1 cell/toxicant mixing ratio may have been the reason that the assays are not sensitive enough because a 9:1 cell solution/toxicant solution ratio often results in a significantly reduced toxicant concentration. Many toxicants are known to possess the ability to adsorb onto the surface of microorganisms. Therefore, a high concentration of cells in the cell-toxicant mixture probably leads to a decrease in the aqueous phase toxicant concentration due to adsorption onto biomass.

The 9:1 cell-toxicant ratio was necessary because all of the new luminescent sensor organisms except PM6 produced light at low levels under the assay conditions. When the activated cell solutions were diluted ten times (as in the protocol described by Lajoie et al., 2002), these organisms (with the exception of PM6) provided light at a level too low for the instrument signal to be significantly different from the background signal. By contrast, PM6 (and the previously created Shk1) produce light that is one to two orders of magnitude higher than that produced by the other sensor organisms. The high light level emitted by PM6 and Shk1 enable assays involving these organisms to be conducted with a 1:9 cell-toxicant ratio, which was determined to be the cell to toxicant mixing ratio in the batch-wise assay protocol described in Chapter 3.0.

A comparison of the results in Chapter 3.0 with the results presented here indicate that although the EC_{50} values for certain compounds (e.g., 4-nitrophenol) did not change much, most compounds have significantly smaller EC_{50} values following the final protocol, and many fewer compounds still have censored toxicity data (EC_{50} reported as less than a certain value). Overall, compared with results shown here, results in Chapter 3.0 indicate that changing the cell-toxicant mixing ratio to 1:9 increased the usefulness of the assays.

B.5 Implications of the Preliminary Screening Protocol

The project team generated toxicity data for metals and organic compounds for PM2, PM5, and PM6, and toxicity data for metals for H1 and H2. HM3 emitted a light level too low to use it in the toxicity assay employed here. The assay procedure involved mixing cell solution with toxicant solution at a ratio of 9:1 (v/v). Many compounds did not reach EC_{50} at the highest concentrations tested (which were much higher than would be realized in a wastewater treatment plant);hence, the sensitivities of the assays were low. As a result, the strains H1, H2, H3, PM2, and PM5 were judged inappropriate for inclusion in a biosensor array for wastewater screening based on the assay procedure used here. In Chapter 3.0, therefore, the project team continued toxicity assays using only the previously created Shk1 and the new strain PM6. These strains produced sufficient light so that the mixing ratio of cells-to-toxicant was decreased to 1:9. The team judged this mixing ratio more favorable because it reduced the potential for adsorption of toxicants to cells.

Table B-1. Toxicity Data (EC_{50}, mg/L) for the Preliminary Toxicity Screening Protocol.

Metals	H1	H2	PM2	PM5	PM6
Cr^{3+}	38.9	50.22	> 438	> 438	59
Cr^{6+}	9.5	11.29	> 26	> 26	10
Cd^{2+}	> 100	> 100	> 100	> 100	> 100
Cu^{2+}	< 2.95	0.72	> 47	15	8.4
Hg^{2+}	0.33	< 0.23	> 1.5	> 1.5	0.3
Pb^{2+}	> 7.5	> 7.5	> 7.5	> 7.5	> 7.5
Ni^{2+}	> 45	> 45	> 45	> 45	> 45
Zn^{2+}	9.9	7.92	20	17	17
Organics					
Benzene			> 70	> 70	> 70
Nitrobenzene			> 179	> 179	> 179
Ether			> 71	> 71	> 71
Ethyl acetate			> 895	> 895	> 895
Cyclohexanone			>947	398	614
1-Butanol			> 810	> 810	
1-Pentanol			> 245	30	> 245
1-Propanol			> 4105	> 4105	
Chloroform			> 450	> 450	> 450
1,1,1-Trichloroethane			> 107	> 107	> 107
Glutaldehyde			>530	>530	77
Methylene chloride			> 1326	> 1326	> 1326
4-Chlorobenzaldehyde			> 100	> 100	> 100
1,2-Dichlorobenzene			> 15	> 15	> 15
Formaldehyde			> 55	> 55	33
2,4-Dinitrotoluene			> 15	> 15	> 15
Bromoform			> 289	> 289	> 289
Benzyl alcohol			> 523	> 523	> 523
p-Xylene			> 13	> 13	> 13
1,2,3-Trichloropropane			> 139	> 139	> 139
1,5-Dichloropentane			> 11	> 11	> 11
1,4-Dichlorobutane			> 114	> 114	> 114
n-Butylamine			> 74	> 74	15
Trichloroethylene			> 117	> 117	> 117
Cyclohexylamine			>173	>173	24
Bromochloromethane			> 996	> 996	> 996
Diethylamine			> 71	> 71	16
Cyclohexane			> 16	> 16	> 16
Octane			> 14	> 14	> 14
n-Amyl acetate			> 44	> 44	> 44
Isobutyl acetate			> 437	> 437	> 437
1,2-Dibromoethane			> 434	> 434	> 434
Methyl ethyl ketone			> 403	> 403	> 403

Ethanolamine	>220	>220	22
m-Cresol	>207	>207	> 207
2-Chlorophenol	> 124	> 124	53
2,3,4,5-Tetrachlorophenol	0.9	> 10	0.7
3-Phenylphenol	13	35	9.2
2,4-Dinitrophenol	> 50	> 50	14
4-Nitrophenol	> 100	> 100	38
1-Hexanol	> 410	> 410	> 410
2,6-Dichlorophenol	> 50	> 50	> 50
4-Chloro-3-methylphenol	>50	>50	26
4-Chlorophenol	> 50	> 50	> 50
2,4-Dichlorophenol	> 10	> 10	> 10
Ethylbenzene	> 13	> 13	> 13
1-Naphthol	> 10	> 10	> 10
2,4,5-Trichlorophenol	1.5	1.5	1.6

REFERENCES

Anderson, S. 1970. *Experimental Microbial Ecology*, Academic Press, New York, p. 129.

Blaise, C., N. Bermingham, and R. Van Collie. 1985. The integrated ecotoxicological approach to assessment of ecotoxicity. *Wat. Qual. Bull.* 10: 3-10.

Blaise, C. 1991. Microbiotests in aquatic ecotoxicology: Characteristics, utility, and prospects. *Environ. Toxicol. Wat. Qual.* 6: 145-155.

Bliss, C.I. 1939. The toxicity of poisons applied jointly. *Ann. Appl. Biol.* 26: 585-615.

Blum, D.J.W. and R.E. Speece. 1991. A database of chemical toxicity to environmental bacteria and its use in interspecies comparisons and correlations. *Res. J. Wat. Pollut. Control Fed.* 63: 198-207.

Bott, C.B. and N.G. Love. 2002. Investigating a mechanistic cause for activated-sludge deflocculation in response to shock loads of toxic electrophilic chemicals, Water Env. Res. 74(3): 306-315.

Burlage, R.S., L.A. Bemis, A.C. Layton, G.S. Sayler, and F. Larimer. 1990. Comparative genetic organization of incompatibility group P degradative plasmids. *J. Bacteriol.* 172(12): 6818-6825.

Castillo, G.C., I.C. Vila, and E. Neild. 2000. Ecotoxicity assessment of metals and wastewater using multitrophic assays. *Environ. Toxicol.* 15: 370-375. Chan, C.M., W.H. Lo, K.Y. Wong, and W.F. Chung. 1999. Monitoring the toxicity of phenolic chemicals to activated sludge using a novel optical scanning respirometer. *Chemosphere* 39: 1421-1432.

Cooper, C. 2001. Use of Microtox for assessing heavy metal complex formation with the organic solvents acetonitrile and dimethyl sulphoxide: a preliminary study. *Bull. Environ. Cont. Toxicol.* 67: 526-534.

De Lorenzo, V., M. Herrero, U. Jakubzik and K.N. Timmis. 1990. Mini-Tn5 transposon derivatives for insertion mutagenesis, promoter probing, and chromosomal insertion of cloned DNA in gram-negative eubacteria. *J. Bacteriol.* 172(11): 6568-6572.

Dutka, B.J. and K.K. Kwan. 1981. Comparison of three microbial toxicity screening tests with the Microtox system. *Bull. Environ. Cont. Tox.*, 27: 753,757.

Dutka, B.J. and K.K. Kwan. 1982. Application of four bacterial screening procedures to assess changes in the toxicity of chemicals in mixtures. *J. Environ. Pollut., Ser. A.*, 29: 125-134.

Dukta, B.J., N. Nyholm, and J. Paterson. 1983. Comparison of several microbiological toxicity screening tests. *Water Res.*, 17: 1363-1368.

Dutka, B.J. and K.K. Kwan. 1984. Studies on a synthetic activated sludge toxicity screening procedure with comparison to three microbial toxicity tests. In D. Liu and B.J. Dutka, eds., *Toxicity Screening Procedures Using Bacterial Systems*, Marcel Dekker, New York, p. 125-138.

Evans, M.R., G.M. Jordinson, D.M. Rawson, and J.G. Rogerson. 1998. Biosensors for the measurement of toxicity of wastewaters to activated sludge. *Pesticide Science* 54:447-452.

Fochtman, P., A. Raszka, and E. Nierzedska. 2000. The use of conventional bioassays, microbiotests, and some "rapid" methods in the selection of an optimal test battery for the assessment of pesticides toxicity. *Environ. Toxicol.* 15: 376-384.

Franklin, N.M., J.L. Stauber, C. Apte, and R.P. Lim. 2002. Effect of initial cell density on the bioavalability and toxicity of copper in microalgal bioassays. *Environ. Toxicol. Chem.* 21: 742-751.

Frei, W. 1913. Versuche über kombinationen von desnfektionsmitteln. *Zeitschrift Hygiene Infektionskrankheiten* 75: 433-496.

Gokcay, C.F. and U. Yetis. 1991. Effect of chromium (VI) on activated sludge. *Water Res.*, 25: 65-73.

Gu, M.B. and S.H. Choi. 2001. Monitoring and classification of toxicity using recombinant bioluminescent bacteria. Water Science and Technology 43:147-154.

Gu, M.B. and G.C. Gil. 2001. A multi-channel continuous toxicity monitoring system using recombinant bioluminescent bacteria for classification of toxicity. *Biosensors & Bioelectronics* 16:661-666.

Hastings, J.W. and K.H. Nealson. 1977. Bacterial bioluminescence. *Annual Review of Microbiology,* 31: 549-595.

Iizumi, T., M. Mizumoto, and K. Nakamura. 1998. A bioluminescence assay using *Nitrosomonas europaea* for rapid and sensitive detection of nitrification inhibitors. *Appl. Environ. Microbiol.* 64(10): 3656-3662.

Kahru, A., L. Põllumaa, R. Reiman, A, Rätsep, M. Liiders, and A. Maloveryan. 2000. The toxicity and biodegradability of eight main phenolic compounds characteristic to the oil-shale industry wastewaters: A test battery approach. *Environ. Toxicol.* 15: 431-442.

Kaiser, K.L.E. and J. Devillers 1994. *Ecotoxicity of Chemicals to Photobacterium phosphoreum. Handbooks of Ecotoxicology Data.* Volume 2. Gordon and Breach Science, Langhorne, Pennsylvania.

Kaiser, S.K., J.B.Guckert, and D.W. Gledhill. 1997. "Comparison of activated sludge microbial communities using biolog microplates." In Proceedings of the Second International Conference on Microorganisms in Activated Sludge and Biofilm Processes. Edited by D. Jenkins and S.W. Hermanowicz. Berkeley, CA.

Lajoie, C.A., S.C. Lin, H. Nguyen, and C.J. Kelly. 2002. A toxicity testing protocol using a bioluminescent reporter bacterium from activated sludge. *J. Microbiol. Met.* 50: 273-282.

Lappalainen, J., R. Juvonen, J. Nurmi, and M. Karp. 2001. Automated color correction method for *Vibrio fischeri* toxicity test. Comparison of standard and kinetic assays. *Chemosphere* 45: 635-641.

Layton, A.C., P.N. Karanth, C.A. Lajoie, A.J. Meyers, I.R. Gregory, R.D. Stapleton, D.E. Taylor, and G.S. Sayler. 2000. Quantification of *Hyphomicrobium* populations in activated sludge from an industrial wastewater treatment system as determined by 16S rRNA analysis. *Appl. Environ. Microbiol.* 66(3):1167-1174.

Madoni, P. 1994. Estimates of ciliated protozoa biomass in activated sludge and biofilm. *Bioresource Technol*., 48**,** 2459-2464.

Mazierski, J. 1995. Effect of chromium (Cr^{VI}) on the growth rate of activated sludge bacteria. *Water Res*., 29: 1479-1482.

Mowat, F. and K.J. Bundy. 2002. Experimental and mathematical/computational assessment of the acute toxicity of chemical mixtures from the Microtox® assay. *Adv. Environ. Res.* 6: 547-558.

Ren, S. and P.D. Frymier. Under revision. Toxicity Screening of Organic Compounds that Inhibit Activated Sludge Microorganisms by Bioluminescent Bioreporter, *Water Environ. Res.*

Ren, S. and P.D. Frymier. 2002. Estimating the toxicities of organic chemicals to bioluminescent bacterium and activated sludge. *Water Res.* 36: 4406-4414.

Ren, S. and P.D. Frymier. 2003a. The use of a genetically engineered *Pseudomonas* species (Shk1) as a bioluminescent reporter for heavy metal toxicity screening in wastewater treatment plant influent. *Water Environ. Res.* 75: 21-29.

Ren, S. and P.D. Frymier. 2003b. Toxicity estimation of phenolic compounds by bioluminescent bacteria. *J. Environ. Eng.* 129: 328-335.

Ren, S. and P.D. Frymier. 2003c. Use of multidimensional scaling in the selection of wastewater toxicity test battery components. *Water Res.* 37: 1655-1661.

Ribo, J. M. 1997. Interlaboratory comparison studies of the luminescent bacteria toxicity bioassay. *Environ. Toxicol. Wat. Qual.* 12: 283-294.

Rouf, M.A. and J.L. Stokes. 1964. Morphology, nutrition and physiology of *Sphaerotilus natans*. *Arch. Microbiol.* 49: 132-149.

Scheers, E.M., C, Van der Wielen, and P. J. Dierickx. 2002. Toxicological evaluation of waste-water samples to appropriately sensitized culture fathead minnow cells compared with the Microtox assay. *Bull. Environ. Cont. Toxicol.* 68: 253-260.

Schultz, T.W., G.D. Sinks, and A.P. Bearden. 1998. QSAR in aquatic toxicology: a mechanism of action approach comparing toxic potency to *Pimephales promelas*, *Tetrahymena pyriformis*, and *Vibrio fischeri*. In J. Devillers, ed., *Comparative QSAR*. Taylor and Francis, New York, NY.

Shaw J.J., P. Rogowsky, T.J. Close, and C.I. Kado. 1987. Working with bacterial bioluminescence. *Plant Molec. Biol. Rep.* 5, 225-236.

Sun, B., N. Nirmalakhandan, E. Hall, X.H. Wang, J. Prakash, and R. Maynes. 1994. Estimating toxicity of organic chemicals to activated-sludge microorganisms. *J. Environ. Eng.* 120, 1459-1469.

Volskay, V.T. and C.P. Leslie Grady. 1988. Toxicity of selected RCRA compounds to activated sludge microorganisms. *J. Water Pollut. Control Fed.* 60, 1850-1856.

Walker, J.D. 1989. Effects of chemicals on microorganisms. *J. Water Pollut. Control Fed.* 61, 1077-1097.

Wang, X., C. Sun, Y. Wang, and L. Wang. 2002. Quantitative structure-activity relationships for the inhibition toxicity to root elongation of *Cucumis sativus* of selected phenols and interspecies correlation with *Tetrahymena pyriformis*. *Chemosphere* 46: 153-161.

Yetis, U. and C.F. Gokcay. 1989. Effect of nickel (II) on activated sludge. Water Res, 23, 10031007.

WASTEWATER UTILITY

Alabama
Montgomery Water Works & Sanitary Sewer Board

Alaska
Anchorage Water & Wastewater Utility

Arizona
Gila Resources
Glendale, City of, Utilities Department
Mesa, City of
Peoria, City of
Phoenix Water Services Department
Pima County Wastewater Management

Arkansas
Little Rock Wastewater Utility

California
Calaveras County Water District
Central Contra Costa Sanitary District
Contra Costa Water District
Crestline Sanitation District
Delta Diablo Sanitation District
Dublin San Ramon Services District
East Bay Dischargers Authority
East Bay Municipal Utility District
El Dorado Irrigation District
Fairfield-Suisun Sewer District
Fresno Department of Public Utilities
Irvine Ranch Water District
Las Virgenes Municipal Water District
Livermore, City of
Lodi, City of
Los Angeles, City of
Los Angeles County, Sanitation Districts of
Napa Sanitation District
Orange County Sanitation District
Palo Alto, City of
Riverside, City of
Sacramento Regional County Sanitation District
San Diego Metropolitan Wastewater Department, City of
San Francisco, City & County of
San Jose, City of
Santa Barbara, City of
Santa Rosa, City of
South Bayside System Authority
South Coast Water District
South Orange County Wastewater Authority
Union Sanitary District
West Valley Sanitation District

Colorado
Boulder, City of
Colorado Springs Utilities
Littleton/Englewood Water Pollution Control Plant
Metro Wastewater Reclamation District, Denver

Connecticut
The Mattabasset District
New Haven, City of, WPCA

District of Columbia
District of Columbia Water & Sewer Authority

Florida
Broward, County of
Fort Lauderdale, City of
Gainesville Regional Utilities
JEA
Miami-Dade Water & Sewer Authority
Orange County Utilities Department
Orlando, City of
Reedy Creek Improvement District
Seminole County Environmental Services
St. Petersburg, City of
Stuart Public Utilities
Tallahassee, City of
Tampa, City of
Toho Water Authority
West Palm Beach, City of

Georgia
Atlanta Department of Watershed Management
Augusta, City of
Clayton County Water Authority
Cobb County Water System
Columbus Water Works
Fulton County
Gwinnett County Department of Public Utilities
Macon Water Authority
Savannah, City of

Hawaii
Honolulu, City and County of

Illinois
American Bottoms Wastewater Treatment Plant
Greater Peoria Sanitary District
Kankakee River Metropolitan Agency
Metropolitan Water Reclamation District of Greater Chicago
North Shore Sanitary District
Wheaton Sanitary District

Indiana
Fort Wayne, City of

Iowa
Ames, City of
Cedar Rapids Wastewater Facility
Des Moines Metro Wastewater Reclamation Authority
Iowa City, City of

Kansas
Johnson County Unified Wastewater Districts
Unified Government of Wyandotte County/Kansas City, City of

Kentucky
Louisville & Jefferson County Metropolitan Sewer District

Louisiana
Sewerage & Water Board of New Orleans

Maine
Bangor, City of
Portland Water District

Maryland
Anne Arundel County Bureau of Utility Operations
Howard County Department of Public Works
Washington Suburban Sanitary Commission

Massachusetts
Boston Water & Sewer Commission
Upper Blackstone Water Pollution Abatement District

Michigan
Ann Arbor, City of
Detroit, City of
Holland Board of Public Works
Lansing, City of
Owosso Mid-Shiawassee County WWTP
Saginaw, City of
Wayne County Department of Environment
Wyoming, City of

Minnesota
Rochester, City of
Western Lake Superior Sanitary District

Missouri
Independence, City of
Kansas City Missouri Water Services Department
Little Blue Valley Sewer District

Nebraska
Lincoln Wastewater System

Nevada
Henderson, City of

New Jersey
Passaic Valley Sewerage Commissioners

New York
New York City Department of Environmental Protection
Rockland County Solid Waste Management Authority/ Sewer District

North Carolina
Cary, Town of
Charlotte/Mecklenburg Utilities
Durham, City of
Metropolitan Sewerage District of Buncombe County
Orange Water & Sewer Authority

Ohio
Akron, City of
Butler County Department of Environmental Services
Columbus, City of
Metropolitan Sewer District of Greater Cincinnati
Northeast Ohio Regional Sewer District
Summit, County of

Oklahoma
Tulsa, City of

Oregon
Clean Water Services
Eugene/Springfield Water Pollution Control
Water Environment Services

Pennsylvania
Philadelphia, City of
University Area Joint Authority, State College

South Carolina
Charleston Commissioners of Public Works
Mount Pleasant Waterworks & Sewer Commission
Spartanburg Sanitary Sewer District

Tennessee
Cleveland, City of
Knoxville Utilities Board
Murfreesboro Water & Sewer Department
Nashville Metro Water Services

Texas
Austin, City of
Dallas Water Utilities
Denton, City of
El Paso Water Utilities
Fort Worth, City of
Gulf Coast Waste Disposal Authority

WERF SUBSCRIBERS

Houston, City of

San Antonio Water System

Trinity River Authority

Utah

Salt Lake City Corporation

Virginia

Alexandria Sanitation Authority

Arlington, County of

Fairfax County Virginia

Hampton Roads Sanitation District

Henrico, County of

Hopewell Regional Wastewater Treatment Facility

Loudoun County Sanitation Authority

Lynchburg Regional WWTP

Prince William County Service Authority

Richmond, City of

Rivanna Water & Sewer Authority

Washington

Edmonds, City of

Everett, City of

King County Department of Natural Resources

Seattle Public Utilities

Sunnyside, Port of

Yakima, City of

Wisconsin

Green Bay Metro Sewerage District

Madison Metropolitan Sewerage District

Milwaukee Metropolitan Sewerage District

Racine, City of

Sheboygan Regional Wastewater Treatment

Wausau Water Works

Australia

South Australian Water Corp.

Sydney Water Corp.

Water Corp. of Western Australia

Canada

Greater Vancouver Regional District

Toronto, City of, Ontario

Winnipeg, City of, Manitoba

Mexico

Servicios de Agua y Drenaje de Monterrey, I.P.D.

New Zealand

Watercare Services Limited

Singapore

Singapore Public Utilities Board

United Kingdom

Yorkshire Water Services Limited

STORMWATER UTILITY

California

Los Angeles, City of, Department of Public Works

Monterey, City of

Santa Rosa, City of

Colorado

Boulder, City of

Georgia

Griffin, City of

Iowa

Cedar Rapids Wastewater Facility

Des Moines Metro Wastewater Reclamation Authority

Kansas

Overland Park, City of

Kentucky

Louisville & Jefferson County Metropolitan Sewer District

Maine

Portland Water District

Minnesota

Western Lake Superior Sanitary District

North Carolina

Cary, Town of

Charlotte, City of, Stormwater Services

Pennsylvania

Philadelphia, City of

Tennessee

Chattanooga Storm Water Management

Washington

Seattle Public Utilities

STATE

Arkansas Department of Environmental Quality

Fresno Metropolitan Flood Control District, Calif.

Urban Drainage & Flood Control District, Colo.

CORPORATE

ADS Environmental Services

The ADVENT Group Inc.

Alan Plummer & Associates

Alden Research Laboratory

Alpine Technology Inc.

Aqua-Aerobic Systems, Inc.

Aquateam–Norwegian Water Technology Centre A/S

BaySaver Inc.

BioVir Laboratories, Inc.

Black & Veatch

Boyle Engineering Corporation

BPR CSO

Brown & Caldwell

Burns & McDonnell

CABE Associates Inc.

The Cadmus Group

Camp Dresser & McKee Inc.

Carollo Engineers Inc.

Carpenter Environmental Associates Inc.

CDS Technologies Inc.

Chemtrac Systems Inc.

CH2M HILL

Damon S. Williams Associates, LLC

David L. Sheridan, P.C.

Earth Tech Inc.

Eco-Matrix

Environmental Engineers International

EMA Inc.

The Eshelman Company Inc.

Finkbeiner, Pettis, & Strout Inc.

Freese & Nichols, Inc.

Frontier Geosciences Inc.

ftn Associates Inc.

Fuss & O'Neill, Inc.

Gannett Fleming Inc.

Golder Associates Inc.

Greeley and Hansen LLC

Hazen & Sawyer, P.C.

HDR Engineering Inc.

HNTB Corporation

HydroQual Inc.

Ingersoll-Rand Energy Systems

Insituform Technologies Inc.

Institute for Environmental Technology & Industry, Korea

Jacobson Helgoth Consultants Inc.

Jason Consultants Inc.

Jordan, Jones, & Goulding Inc.

KCI Technologies Inc.

Kelly & Weaver, P.C.

Kennedy/Jenks Consultants

Komline Sanderson Engineering Corporation

Lawler, Matusky & Skelly Engineers, LLP

Limno-Tech Inc.

Lombardo Associates Inc.

Malcolm Pirnie Inc.

McKim & Creed

MEC Analytical Systems Inc.

Metcalf & Eddy Inc.

MWH

New England Organics

Odor & Corrosion Technology Consultants Inc. (OCTC)

ONDEO Degremont Inc.

Oswald Green, LLC

PA Government Services Inc.

Parametrix Inc.

Parsons

Post, Buckley, Schuh & Jernigan

R&D Engineering/Conestoga Rover & Associates

The RETEC Group

RMC, Inc. (Raines, Melton & Carella)

R.M. Towill Corporation

Ross & Associates Ltd.

Royce Technologies

SAIC Maritime Technical Group

Sear-Brown

Short Elliott Hendrickson, Inc.

Stormwater Management, Inc.

Synagro Technologies, Inc.

Tetra Tech Inc.

Trojan Technologies Inc.

URS Corporation

Veolia Water NATC

Wade-Trim Inc.

Weston Solutions Inc.

Woodard & Curran

WRc/D&B, LLC

WWETCO, LLC

Zoeller Pump Company

INDUSTRY

American Electric Power

ChevronTexaco Energy Research & Technology Company

The Coca-Cola Company

Dow Chemical Company

DuPont Company

Eastman Kodak Company

Merck & Company Inc.

ONDEO Services

Procter & Gamble Company

PSEG Services Corp.

Severn Trent Services Inc.

Shell Global Solutions

Thames Water Plc

United Water Services LLC

Note: List as of 1/31/04

WERF Product Order Form

As a benefit of joining the Water Environment Research Foundation, subscribers are entitled to receive one complimentary copy of all final reports and other products. Additional copies are available at cost (usually $10). To order your complimentary copy of a report, please write "free" in the unit price column. WERF keeps track of all orders. If the charge differs from what is shown here, we will call to confirm the total before processing.

Name _______________ Title _______________

Organization _______________

Address _______________

City _______________ State _______ Zip Code _______ Country _______

Phone _______________ Fax _______________ Email _______________

Stock #	Product	Quantity	Unit Price	Total
			Postage & Handling	
			VA Residents Add 4.5% Sales Tax	
			Canadian Residents Add 7% GST	
			TOTAL	

Method of Payment:

(All orders must be prepaid.)

❑ Check or Money Order Enclosed

❑ Visa ❑ Mastercard ❑ American Express

Account No. _______________ Exp. Date _______

Signature _______________

Shipping & Handling:

Amount of Order	United States	Canada & Mexico	All Others
Up to but not more than:	Add:	Add:	Add:
$20.00	$5.00*	$8.00	50% of amount
30.00	5.50	8.00	40% of amount
40.00	6.00	8.00	
50.00	6.50	14.00	
60.00	7.00	14.00	
80.00	8.00	14.00	
100.00	10.00	21.00	
150.00	12.50	28.00	
200.00	15.00	35.00	
More than $200.00	Add 20% of order	Add 20% of order	

*minimum amount for all orders

Note: Please make checks payable to the Water Environment Research Foundation.

To Order (Subscribers Only):

Log on to www.werf.org and click on the "Product Catalog."

Phone: (703) 684-2470
Fax: (703) 299-0742.

WERF
Attn: Subscriber Services
635 Slaters Lane
Alexandria, VA 22314-1177

To Order (Non-Subscribers):

Non-subscribers may be able to order WERF publications either through WEF (www.wef.org) or IWAP (www.iwapublishing.com). Visit WERF's website at www.werf.org for details.